SPINOFF 1995

National Aeronautics and Space Administration
Office of Space Access and Technology
Commercial Development and Technology Transfer Division

By James J. Haggerty

National Aeronautics and
Space Administration

INTRODUCTION

President Clinton has called technology "the engine of economic growth." It is a key element in attaining one of the major goals enunciated by the President: reinvigorating American competitiveness in the global marketplace.

Technological innovation is today the principal currency of international competition. The rewards for bringing superior new products and processes to the marketplace are a robust economy for the successful nation and an elevated standard of living for its people.

The great storehouse of technology NASA has built in 37 years of pursuing aeronautical and space goals represents a singularly important asset for the United States. That technology is not perishable; it can be used over and over, and it can be adapted to uses different and often remote from the original application.

This bank of knowledge is, in fact, being well utilized. Hundreds of companies have taken advantage of this national resource to develop more than a thousand new products and processes for the world market. They gain a competitive edge in the marketplace because they employ advanced technology already developed.

These secondary applications of aerospace technology — "spinoffs" — span so broad a range of public needs and conveniences it is almost impossible to find an area of everyday life they have not improved. Collectively they represent a substantial dividend on the national investment in aerospace research.

Recognizing the great potential of the technology bank, Congress charged NASA with stimulating the widest possible use of this valuable resource in the national interest. NASA's instrument of that purpose is the Technology Transfer Program, which seeks to broaden and accelerate the spinoff process. Its intent is to spur expanded national benefit, in terms of new products and new jobs, by facilitating the commercial application of the technology; it encourages greater use of the storehouse of knowledge by providing a channel linking the technology and those who might be able to put it to advantageous use. In July 1994, NASA implemented an Agenda for Change — a new way of doing business in partnership with the private sector. This Agenda marks the beginning of a new focus to further improve our contributions to Amercia's economic security through the pursuit of aeronautics and space missions.

This publication is an implement of the Technology Transfer Program intended to heighten awareness among potential users of the technology available for transfer and the economic and social benefits that might be realized by applications of NASA technology to U.S. commercial interests.

Spinoff 1995 is organized in three sections:

Section 1, outlines NASA's mainline effort, the major programs that generate new technology and therefore replenish and expand the bank of technical knowledge available for application.

Section 2, the focal point of this volume, contains a representative sampling of spinoff products and processes that resulted from applications of technology originally developed to meet NASA aerospace goals.

Section 3, describes the various mechanisms NASA employs to stimulate technology transfer and lists, in an appendix, contact sources for further information about the Technology Transfer Program.

Dr. Robert L. Norwood
Director, Commercial Development and Technology Transfer Division
National Aeronautics and Space Administration

FOREWORD

On July 4, 1995, American astronauts and Russian cosmonauts concluded five days of joint orbital research in their linked spacecraft. The rendezvous and docking of the Space Shuttle Atlantis and the Mir space station was a historic mission, the beginning of a new era of expanded international cooperation.

For the United States, it was something more, the 100th U.S. human space flight. That milestone recalls memories of the first Americans in space in 1961 and 1962; the first spacewalks and rendezvous/docking missions of the '60s; the momentous years of 1969-72 when a dozen Americans walked on the Moon and firmly reestablished U.S. pre-eminence in space; and some 70 flights of the Space Shuttle that have demonstrated an extraordinary capability for flexible, highly sophisticated operations in Earth orbit.

These 100 flights, over 34 years, have provided a solid foundation of experience and technology for the next major step in human space activity, the assembly and operation of the International Space Station. The Space Station is vital to the future of space research, but its importance goes well beyond that: It is a facility where scientists can take advantage of the microgravity environment to conduct experiments that promise dramatic advances in materials and medications used on Earth for combating intractable diseases.

The International Space Station, with the partnership of Canada, Europe, Japan, Russia, and the United States, represents only one segment of NASA's comprehensive program of research. We are engaged in a broad space science effort that will provide vast new knowledge about the universe and Earth's place in it. That knowledge is immensely valuable in its own right and creates a foundation to provide for tomorrow's practical applications.

We are continuing our Mission to Planet Earth, which seeks greater understanding of the complex mechanisms that control our planet's behavior and application of that knowledge for the benefit of all on Earth.

We are developing cutting-edge aeronautical technologies with our partners in government, industry, and academia that will help elevate the competitive capability of the U.S. aeronautics industries and contribute to development of safer, more efficient, and more environmentally compatible aircraft, while developing the advanced technologies for a High Speed Civil Transport, reducing air traffic delays, enhancing the safety and effectiveness of the national airspace system, and revitalizing the general aviation industry.

While pursuing these multiple responsibilities, we are also conducting a revolutionary restructuring of NASA's organization and operations designed to effect large-scale economies. We are adapting to the realities of the times and making fundamental changes in our methods, workforce, facilities, and relationships with contractors — everything that promises reduced outlays without significant loss of capability.

There will, however, be no change in the vision with which we establish goals nor the dedication with which we pursue them. I am confident that, with the support of the American people, we can meet the challenge of doing more with less, and that we can carry out an effective program that will ensure continued U.S. leadership in aeronautical and space research.

Daniel S. Goldin
Administrator
National Aeronautics and Space Administration

CONTENTS

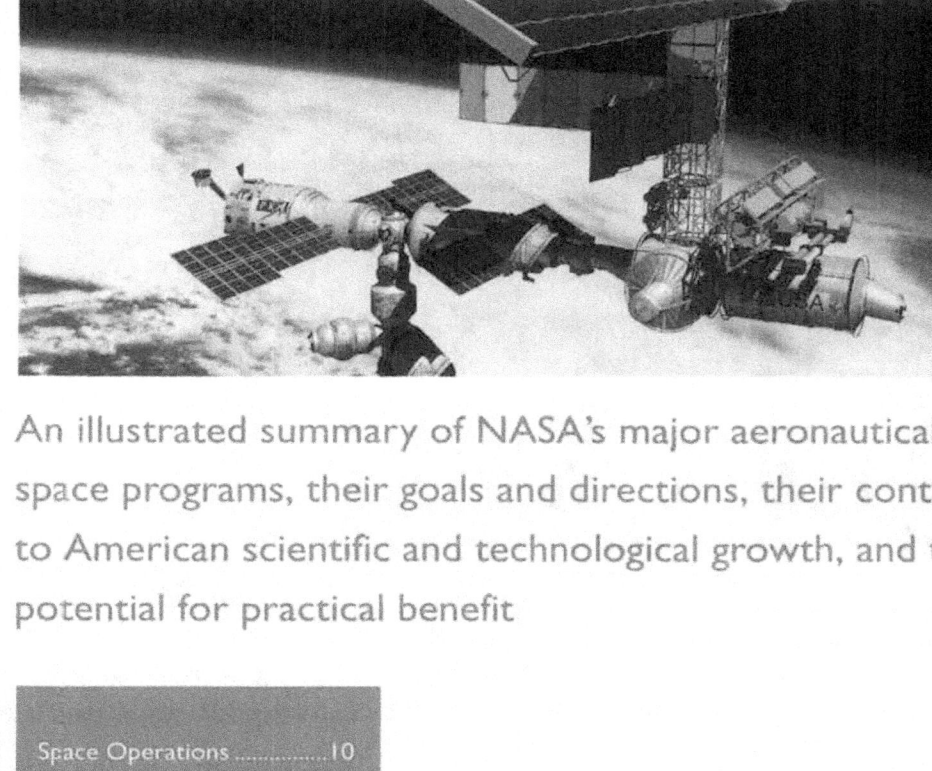

An illustrated summary of NASA's major aeronautical and space programs, their goals and directions, their contributions to American scientific and technological growth, and their potential for practical benefit

AEROSPACE AIMS

U.S./Russian coopera-
tive missions highlight
the stepping stones to
establishment of the
International Space
Station

On July 7, 1995, the Space Shuttle Orbiter *Atlantis* touched down at Kennedy Space Center, concluding one of the most dramatic, interest-capturing space missions since the Apollo era. Executed with textbook precision, the historic STS-71 flight featured a challenging docking with the Russian space station *Mir* and five days of cooperative Russian/American experimentation in the linked *Atlantis/Mir*, the largest structure ever assembled in orbit.

The mission to *Mir*, the 100th U.S. human space flight, marked a milestone in a multistep preliminary agenda intended to provide a framework for International Space Station assembly operations, which began in 1997.

Phase I of the developmental flight program started in February 1994, when Russian cosmonaut Sergei Krikalev flew aboard the eight-day Shuttle mission STS-60. A year later, on February 3, 1995, NASA launched STS-63, Orbiter *Discovery* to a rendezvous with *Mir*. *Discovery* successfully accomplished the rendezvous, then executed a slow "flyaround," circling the station at a distance of about 450 feet for photography and communications tests. The rendezvous and flyaround validated a number of techniques for subsequent employment in docking missions, such as those essential to assembly of the International Space Station.

On March 14, 1995, U.S. astronaut/physician Norman E. Thagard, together with cosmonauts Vladimir Dezhurov and Gennady Strekalev, were launched to *Mir* to begin a 115-day stay in space conducting life sciences and station-related experiments and preparing *Mir* for its meeting with *Atlantis*.

The historic *Atlantis/Mir* docking occurred on June 29, 161 miles above Central Asia near the Russian/Mongolian border. On board *Atlantis* were commander Robert "Hoot" Gibson; pilot Charles J. Precourt, and mission specialists Bonnie Dubbar, Ellen S. Baker and Gregory J. Harbaugh; they were joined by the two-man replacement crew for *Mir*, cosmonauts Anatoly Solovyev and Nikolia Budarin.

After two days of gradually narrowing the distance between *Atlantis* and *Mir*, Gibson flew the Orbiter to a position 1,000 feet below *Mir*, then used the Orbiter's computer-directed thrusters to close the gap. Precisely on schedule at 1300 hours Greenwich Mean Time, Gibson maneuvered the Orbiter so that its capture ring latched into an interface ring in *Mir's* docking module. A final short burst of the Orbiter's thrusters forced the two spacecraft tightly together in "hard capture." The seven-person crew of *Atlantis* left the Orbiter via an airlock system and floated through a tunnel to the *Mir* core module, where they joined Thagard, Dezhurov and Strekalev — a space "first" with 10 people in orbit at one time.

For the next four days, the international crew conducted a variety of experiments, centering on medical investigations of the effects of long duration space flight on Thagard and his *Mir* crewmates. On the Fourth of July, the docking interfaces were released and *Atlantis* separated from *Mir*, now carrying a somewhat different eight-person crew; while Solovyev and Budarin remained on *Mir*, Thagard, Dezhurov and Strekalev joined the original *Atlantis* team for the return to Florida.

The mission was the first of seven planned Shuttle/*Mir* flights in 1995-97 intended to provide joint flight experience, reduce the risks associated with space station assembly, and allow extensive cooperative research. The demonstration of U.S./Russian teamwork in orbit was, in the words of NASA Administrator Daniel S. Goldin, "a wonderful dream come true," an impressive reinforcement of confidence that the challenging task of assembling the International Space Station will be accomplished with equivalent precision.

(Continued)

This artist's conception depicts the historic docking of the U.S. Shuttle Orbiter Atlantis with the Russian space station Mir. Atlantis is linked to the Kristall module of the six segment Mir by a Russian-developed docking system and a U.S.-developed external airlock.

INTERNATIONAL SPACE STATION

The International Space Station will be a permanent laboratory for human-monitored long term research in the unique environment of Earth-orbital space, an environment that cannot be replicated on Earth for long duration experiments. Research at the station will focus on two key areas: life sciences and materials sciences.

Life sciences research is expected to lead to a clearer understanding of basic processes and that will provide a foundation for development of advanced medications for improved human health. Materials research offers promise of improved metals, composites and plastics for significant advances in technologies for communications, transportation and a broad range of industrial processing operations.

The International Space Station draws upon the resources and scientific/technological expertise of 13 cooperating nations, including the U.S., Canada, Japan, Russia and nine nations of the European Space Agency (Belgium, Denmark, France, Germany, Italy, Norway, Spain, the Netherlands and the United Kingdom). The prime contractor is The Boeing Company and the principal subcontractors are McDonnell Douglas Corporation and the Rocketdyne Division of Rockwell International.

While the Phase I flight program advanced to orbital docking operations in 1995, manufacturers were turning out the first hardware components of the space station. By January 1995, 25,000 pounds of station hardware had been built; it was estimated that the total would reach 100,000 pounds by the end of 1995.

Shown **below** is a major piece of flight hardware rolled out in April 1995: a large Boeing-built module known as a node that will serve as a storage area, a housing for electrical power conversion equipment and as a connecting passageway between the living and laboratory quarters of the space station. The Node 2 pictured will be used for pressure and leak tests and join the space station complement in 1999. In the meantime, Boeing will put the finishing touches to Node 1, whose exterior structure was completed in July 1995; Node 1 will be the first U.S. hardware delivered to orbit. Boeing is building a third node.

Phase II of the space station construction program will begin in November 1997 with the launch on a Russian Proton vehicle of the "FGB." The FGB is a 42,600-pound element, built in Russia but purchased by the U.S., that includes the energy block,

contingency fuel storage, propulsion and multiple docking points.

A month later the Boeing node will be delivered by the Space Shuttle and attached to the FGB. In May 1998, the embryo space station will grow with the addition of the Proton-boosted Russian Service Module, which provides life support and habitation facilities, utilities and thrusters. Shortly thereafter, the crew transfer vehicle — a Russian Soyuz TM capsule — will be joined to the station. At this point, the station will be able to accommodate long duration stays of three-person crews.

Further additions to the expanding station in the latter part of 1998 and early 1999 will include one of the four U.S. solar array modules, which will provide about 23 kilowatts of power; the U.S. laboratory module; the Canadian-built mobile servicing system; and a utilization flight bringing up equipment for outfitting the laboratory module.

In April 1999, Phase II will end with the completion of assembly and outfitting of the interim human-tended configuration of the International Space Station, shown **above.** The large segment at the top is the 112-foot-long U.S. solar array module. The FGB is in the center of the photo; docked with it at far left is the Russian Service Module and below it the Soyuz TM crew transfer vehicle. At the right end of the station is the U.S. laboratory module and Node 1, a storage module. Atop the laboratory is the Canadian mobile servicing system.

(Continued)

In Phase II, the International Space Station will progress gradually from human-tended capability to its ultimate status as a fully operational permanent orbital research facility. Among key additions to the human-tended core configuration are the remaining modules of the U.S. solar array; the Japanese Experiment Module (JEM), to be delivered in March 2000; the European Space Agency (ESA) laboratory, February 2001; and the U.S. habitation module, February 2002. In June 2002, station assembly will be completed with the addition of a second crew transfer vehicle, which will provide an emergency evacuation capability for a full crew of six.

The completed station, shown **below**, will measure 361 feet from tip to tip of the solar arrays. That corresponds to the length of a football field, including both end zones; however, the *area* covered by the station complex is equal to that of two football fields. The pressurized living and working space is roughly equivalent to the passenger cabin volume of two 747 jetliners.

There will be seven laboratories. The U.S. is providing two of them, the basic laboratory module and a Centrifuge Accommodation Module. There will be three Russian research modules, the Japanese JEM and the ESA's *Columbus* Orbital Facility. The U.S., ESA and Japanese laboratories together provide 33 International Standard Payload Racks; additional payload space will be available in the Russian modules. In addition, the JEM has an exposed "back porch," with 10 mounting spaces for experiments that require long duration contact with the space environment; the JEM has a small robotic arm for mounting and moving back porch payloads.

A closer view of the station (**right**) shows the locations of the major segments. In center photo, the U.S.

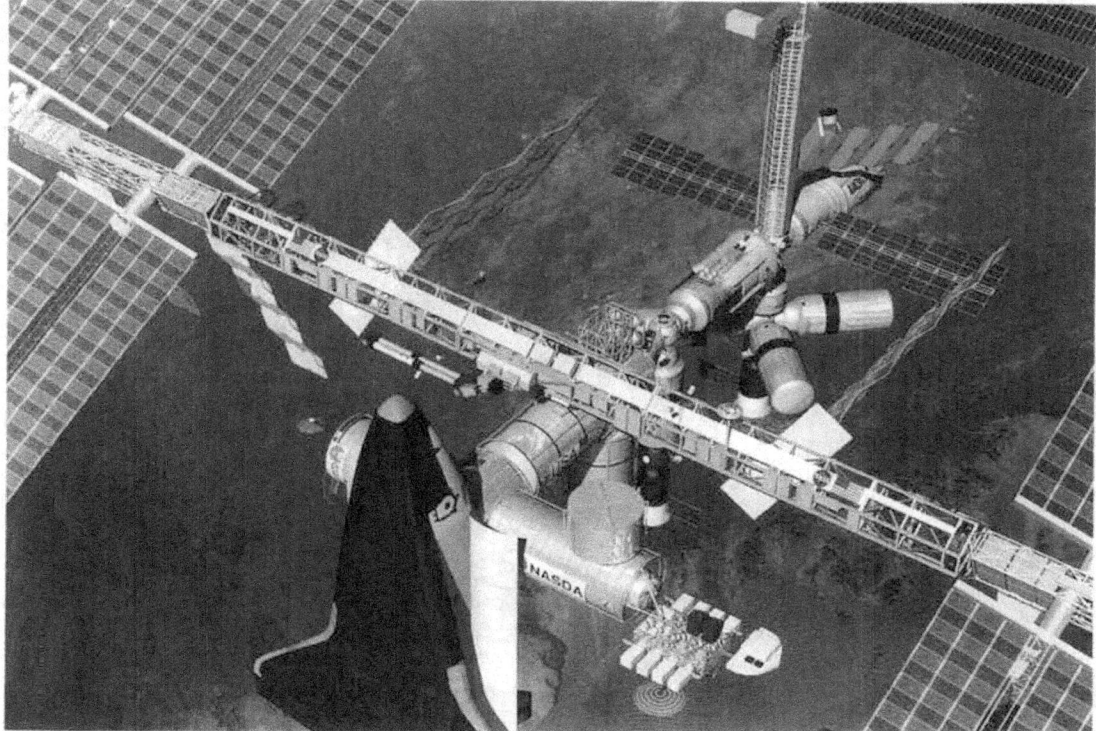

laboratory module is partially hidden by the docked Orbiter; below it is the habitation module, which contains the galley, toilet, shower, sleep stations and medical facilities.

The ESA module is visible to the left of the Orbiter, the JEM (marked NASDA) to the right. The central structure connecting the modules and the main solar power array is the U.S.-built integrated truss; generating a combined 92 kilowatts, the four modules of the power array rotate on the truss, maximizing their exposure to the Sun.

At the "rear" of the station assembly is the Russian service module. The vertical structure next to the service module is the Russian-built Science Power Platform (SPP) with its own solar array; the SPP provides additional power and heat rejection for science and operations. Just above the Orbiter's node, on the integrated truss, is the Canadian module servicing system with its 55-foot robot arm and a mobile transporter that enables it to move along the truss for robotic assembly and maintenance operations.

The International Space Station will operate at an altitude of 240 miles. At that altitude, minute drag forces will cause the station to lose height very gradually, so it will be necessary to reboost it every 90 days. The reboosting will be accomplished by the FGB.

Beginning with the 1997 launches of the Russian Proton and the Space Shuttle, there will be 73 flights until fully operational status is attained in June 2002. The Space Shuttle will make 27 flights, 21 for assembly operations and six for utilization/outfitting. The Russian Proton and Progress, and Ukrainian Zenit launch vehicles will make 45 flights; 19 of them will be made by the Progress vehicle, bringing up propellant for the FGB's reboosting work. There will also be one flight of the Ariane 5 launch vehicle for delivery of ESA's *Columbus* module.

SPACE SHUTTLE OPERATIONS

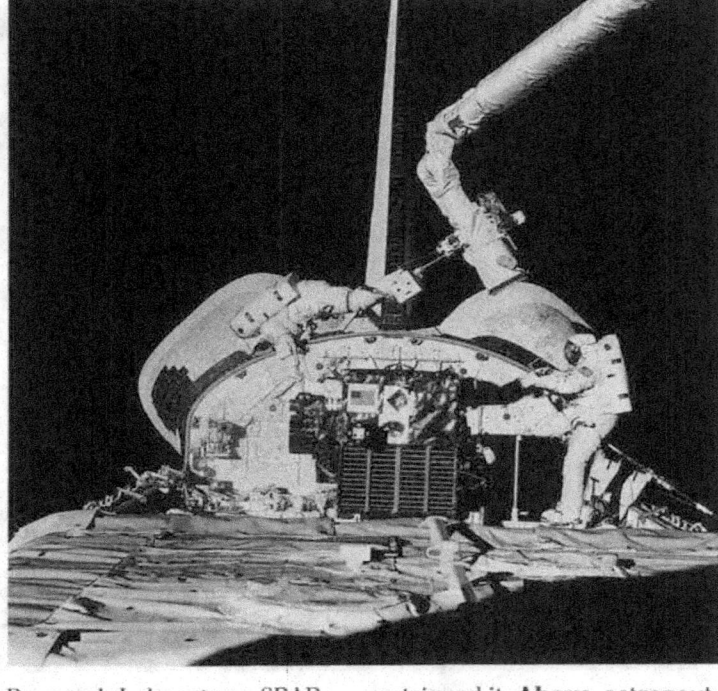

Launched February 3, 1995, flight STS-63, Orbiter *Discovery*, was highlighted by the first rendezvous with the Russian *Mir* space station (see page 10), but it also accomplished a number of other objectives. Among them were the third flight of the SPACEHAB module, a commercially developed space research laboratory located in the forward end of *Discovery's* payload bay (see page 50); and the operation of a free-flying astronomical research satellite, SPARTAN-204.

The STS-63 crew was commanded by James D. Wetherbee with Eileen M. Collins as pilot (she became the first woman to serve as pilot on a Shuttle mission). Also on board were four mission specialists: payload commander Bernard A. Harris Jr., Michael C. Foale, Janice Voss and cosmonaut Vladimir Titov, the second Russian to fly on the Shuttle as part of the *Shuttle/Mir* rendezvous and docking program.

SPARTAN stands for Shuttle Pointed Autonomous Research Tool for Astronomy; it is an observatory designed to obtain data in the far ultraviolet region of the spectrum from diffuse sources of light. Developed by Goddard Space Flight Center and the U.S. Naval

Research Laboratory, SPARTAN is intended to provide easy and relatively inexpensive access to orbit for science experiments.

On STS-63, the SPARTAN mission was conducted in two phases. In the first, the crew used *Discovery's* robot arm to remove the SPARTAN spacecraft from its support structure in the payload bay, then pointed it at the Orbiter's tail to observe surface glow, and pointed it at a thruster to obtain far ultraviolet spectrographs of a thruster firing.

Upon completion of the rendezvous with *Mir*, the crew used the robot arm again to release SPARTAN-204 over the side of the Orbiter to operate independently in space. It flew for 40 hours, observing a variety of celestial targets, then *Discovery* maneuvered t a rendezvous with the satellite and

retrieved it. **Above**, astronaut Foale (in center photo) prepares to grab SPARTAN with the help of payload commander Harris (right).

SPACEHAB-3 carried more than 20 experiments, among them an ASTROCULTURE payload for investigation of growing food plants for long duration space missions; examination of biological activity in microgravity by the BioServe Pilot Laboratory; research on carbohydrate-rich plants as space food; a test of "Charlotte," an experimental robotic device built by McDonnell Douglas Aerospace to demonstrate automated servicing of payloads; a microgravity examination of plant cell shapes and structures; additional research in protein crystal growth; and a demonstration of hardware for controlled liquid phase sintering experiments.

In addition to the SPACE-

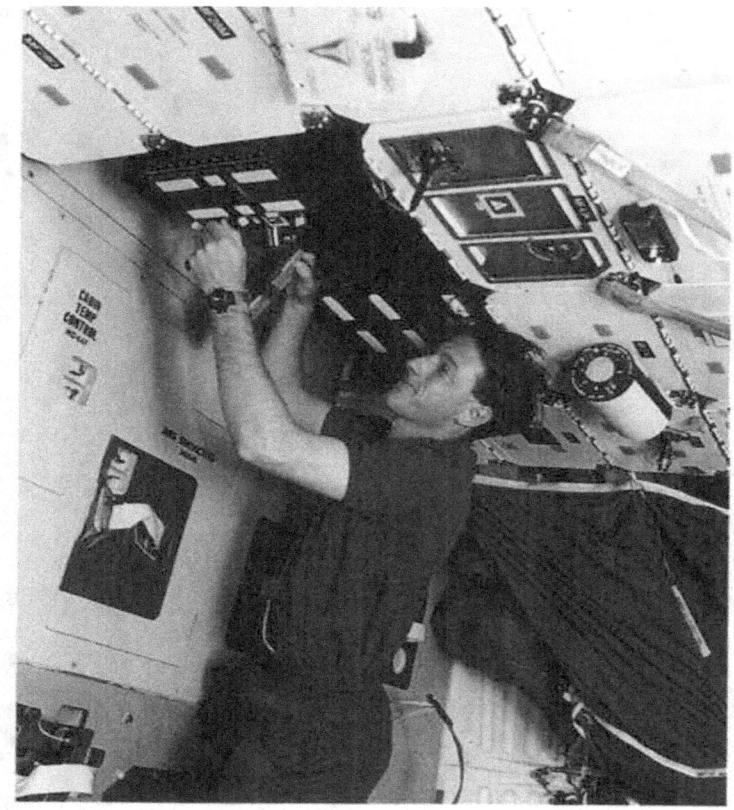

HAB investigations, other research conducted on STS-63 included the first of four Hitchhiker missions planned for 1995; Hitchhiker is a support system that provides power, data and command services to operate the experiments, STS-63 also conducted another test (the eighth) of the Solid Surface Combustion Experiment (SSCE), a major study of how flames spread in microgravity designed to improve knowledge of how flames spread on Earth. **At right,** astronaut Foale is checking the SSCE. **Below,** astronauts Harris and Voss are working a life sciences experiment; Harris, a physician, is checking Voss' biomedical harness to determine the response of muscles to gravity.

The extravehicular activity (EVA) performed by Harris and Foale was also experimental in nature. Their deployment and recapture of the SPARTAN satellite was an exercise in mass handling to increase NASA's experience base in preparation for orbital assembly of the International Space Station. They also evaluated modifications to the EVA suits designed to improve thermal protection from cold.

(Continued)

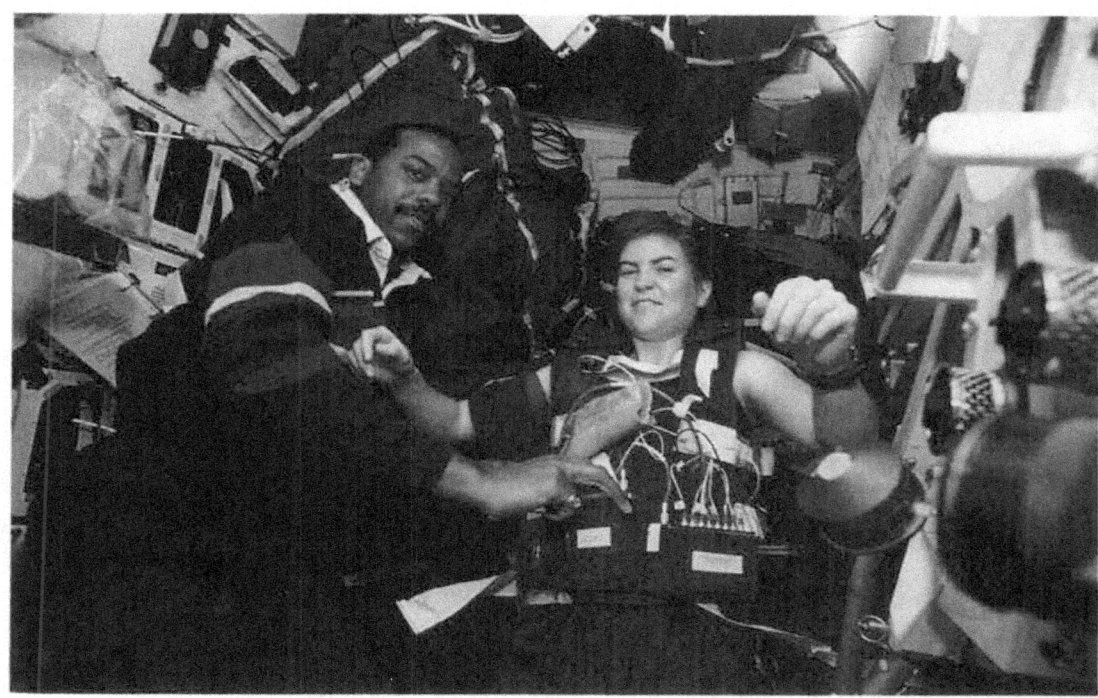

The longest Shuttle mission ever flown began on March 2, 1995 with the launch of STS-67, Orbiter *Endeavour*. The mission, whose principal payload was the Astro-2 observation platform, lasted 17 days.

The crew of seven included five NASA astronauts and two civilian scientists. Commander of STS-67 was astronaut Stephen S. Oswald; the NASA crew included pilot William G. Gregory, payload commander Tamara E. Jernigan and mission specialists John M. Grunsfeld and Wendy B. Lawrence. The civilian payload specialists were Samuel T. Durrance, an astrophysicist with Johns Hopkins University and Ronald A. Parise of Computer Sciences Corporation.

The Astro observatory is a Shuttle-based astronomical system, mounted in the Orbiter's payload bay on two unpressurized Spacelab pallets. Its instruments include three ultraviolet (UV) telescopes aligned to each other on a single pointing instrument; the Johns Hopkins Ultraviolet Telescope (HUT); the University of Wisconsin's UV Photo-Polarimeter Experiment (WUPPE); and Goddard Space Flight Center's UV Imaging Telescope (UIT). Astro-2 marked the second flight of the system; the first was flown in 1990.

Working in two teams, the crew supported round-the-clock Astro-2 observations of UV sources. Typically, the commander or pilot maneuvered *Endeavour* to point

Astro in the general direction of a target source; a mission specialist acquired guide stars and locked Astro into an optical hold on one of these stars; and a payload specialist operated the system's instruments during the observation. Several hundred sources were examined. **At left**, payload commander Jernigan is studying a display of data from the WUPPE experiment.

In addition to Astro, STS-67 carried several experiment packages in the Orbiter's pressurized mid-deck area, among them a protein crystal growth experiment, a Shuttle amateur radio experiment, and the Commercial Materials Dispersion Apparatus Technology Associates Experi-

ment (CMIX-03). **At left,** astronaut Gregory is working on a pharmaceutical experiment in the middeck locker area. CMIX-03 included not only pharmaceutical but also biotechnology, cell biology, fluids and crystal growth investigations intended to explore ways in which microgravity can advance drug development and delivery for treatment of cancer, infectious diseases and metabolic deficiencies.

Another in-cabin experiment was the Middeck Active Control Experiment (MACE), a test platform designed to explore the effects of vibration on spacecraft instruments and structures in microgravity. **Above,** mission commander Oswald is monitoring the experiment; the MACE test platform, shown in upper foreground, is a five-foot composite beam with movable, simulated payloads

at either end. During the mission, Oswald completed several hundred preprogrammed MACE tests.

Among other STS-67 research was a laptop-based Shuttle landing simulator to refresh flight crew experience during long duration missions; a communications experiment investigating the capabilities of the Tracking and Data Relay System (TDRS) with an eye toward eliminating some costly ground stations; and a comparison test of *Endeavour's* own navigation data with data supplied by the Navstar Global Positioning System constellation of satellites.

STS-67 was the second of eight Shuttle missions planned for 1995. The third was the historic STS-71 *Atlantis/Mir* docking (see page 10); *Atlantis* was to make a second docking with *Mir* in November.

The year's fourth mission was STS-70, a flight to deploy a new TDRS satellite that was in progress at publication time. Set for August launch was STS-69, a reflight of the Wake Shield Facility (see page 00) and a test of extravehicular activities related to assembly of the International Space Station. In September, STS-73 was to conduct a second series of U.S. Microgravity Laboratory experiments. The primary assignment of STS-72, targeted for December, was retrieval of a Japanese satellite.

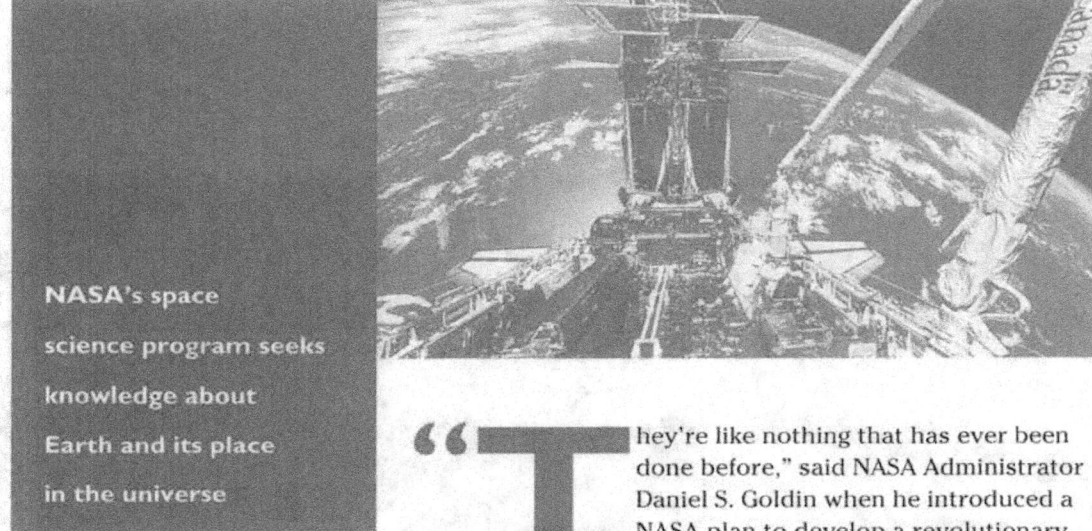

NASA's space
science program seeks
knowledge about
Earth and its place
in the universe

"They're like nothing that has ever been done before," said NASA Administrator Daniel S. Goldin when he introduced a NASA plan to develop a revolutionary series of New Millennium spacecraft. "These small, agile spacecraft will be built on a bench, not in a high-bay area. They'll be about a tenth of the cost and about a tenth of the weight of today's spacecraft."

The New Millennium initiative is intended as a way of allowing infusion of the most advanced technologies into future science missions in an era of ever-tightening federal budgets. It is described as "an aggressive flight technology demonstration effort with simultaneous objectives of increasing the capability of NASA spacecraft, their instruments and their flight rates, while significantly reducing their mass and life cycle costs."

New Millennium spacecraft are designed to permit high-risk orbital testing of untried new technologies at minimal cost; once the technologies are proved in flight, they can be incorporated in operational spacecraft for planetary exploration, Earth observation, astronomy or space physics research.

This initiative exemplifies a broader NASA drive to cut costs across the whole spectrum of space activity, from system design to manufacture to launch to operation and data acquisition — but to do so without sacrificing spacecraft capability. The New Millennium philosophy envisions a series of spacecraft that are not only smaller, lighter and significantly less costly, but also capable of improved performance through use of miniaturized advanced sensors, electronics and components.

Simplified design and new manufacturing techniques will lower the cost of the spacecraft itself. Its reduced size and weight will permit launch by launch vehicles that consume less fuel and thereby provide additional cost savings. And because the spacecraft will be highly autonomous, it can be operated by a smaller ground control team, cutting the cost of operations.

Under the plan, NASA will work in partnership with industry contractors, government laboratories and academia to assure the broadest application of advanced technologies to the New Millennium program. As an initial step, Jet Propulsion Laboratory has embarked on a survey to identify advances in microminiaturization that might be adapted to the new series of "microspacecraft." A tentative schedule envisions the first New Millennium launch before the end of the decade, with follow-on missions at the rate of one every 12-24 months.

The New Millennium proposal initially contemplates technology demonstrations in orbit, but it could produce a new family of operational scientific spacecraft. NASA is currently operating more than a score of science-seeking spacecraft, independently or in concert with other nations. The broad space science program embraces six major areas of activity:

Symbolic of NASA's space science program is this Hubble Space Telescope true color image of Saturn, showing a massive storm (white clouds) stretching 8,000 miles near the planet's equator.

- Solar system exploration, which involves investigations of the planets, moons, comets and asteroids of the solar system.

- Astronomy and astrophysics, which embrace study of the stars and galaxies toward an understanding of the origin and evolution of the universe.

- Space physics, which embraces investigation of the Sun/Earth system, including the study of plasmas and ionized gases originating in the solar system and beyond.

- The Mission To Planet Earth, which seeks understanding of the total Earth system and the effects of natural and human-induced changes in the global environment, and use of the knowledge gained to preserve and improve Earth's environment for future generations.

- Life and biomedical sciences, a program dedicated to advancing medical and biological knowledge toward protecting the health of humans in space, and to the utilization of that knowledge in beneficial Earth applications.

- Microgravity sciences and applications, involving investigation directed toward greater understanding of the airless, weightless Earth orbital environment and its effects on Earth-use materials.

    Examples of current and planned space science activity are contained on the following pages.

SOLAR SYSTEM EXPLORATION

T he historic Ulysses spacecraft, which in 1994 became the first space exploration vehicle to reach a polar region of the Sun, continued its history-making odyssey in 1995 as it passed its perihelion, its closest approach to the Sun.

The spacecraft **(below)**, built by the European Space Agency (ESA) for the joint ESA/NASA mission to study the solar polar areas, passed over the vicinity of the south pole in June 1994 after a four-year journey from Earth. Then, traveling at a velocity of about 73,000 miles per hour with respect to the Sun, Ulysses moved northward to collect data on the Sun's equatorial region, where it reached perihelion on March 12, 1995. At that point the spacecraft was within 124 million miles of the Sun.

In April 1995, the spacecraft crossed into the northern hemisphere of the heliosphere (the region of space dominated by the forces of the solar wind) and in June reached a point 70 degrees north of the Sun's equator. That marked the start of a new phase of the Ulysses mission, a four-month observation of the north polar region. Upon completion of the north polar reconnaissance, Ulysses concluded its primary mission; depending on available funding, it could continue to collect data through part or all of an 11-year solar cycle. The Ulysses program is managed for NASA's Office of Space Science by Jet Propulsion Laboratory (JPL).

In July 1995, the Galileo spacecraft **(right)** began the "final approach" — a 50-million-mile approach — of its six-year roundabout journey to Jupiter, after releasing an instrumented probe that was

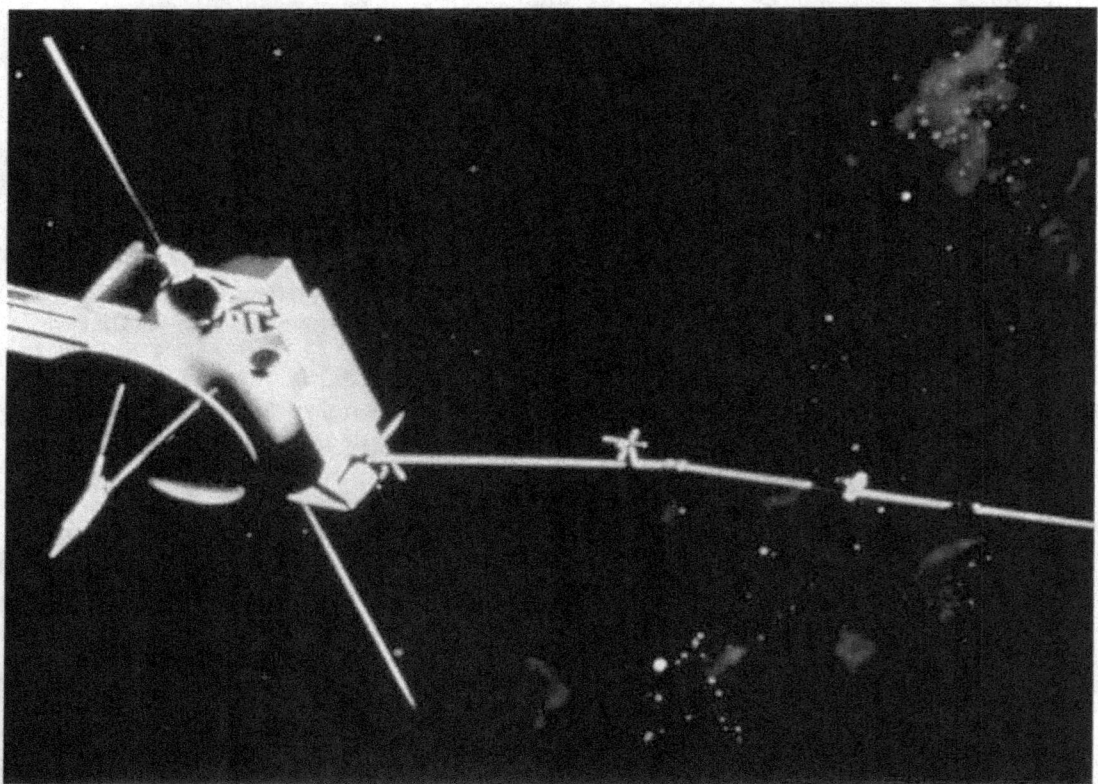

to proceed independently to Jupiter and descend into the Jovian atmosphere. The probe was targeted for arrival at Jupiter on December 7, 1995, at which time it would make the most difficult planetary entry ever attempted. Flying at a speed of 106,000 miles per hour, the probe was expected to encounter deceleration forces up to 350G (350 times Earth's gravity) on its descent some 400 miles into the planet's complex atmosphere, making the first direct measurements of Jupiter's clouds, lightning, winds and other features. The probe was to radio its data to the main Galileo spacecraft for later transmission to Earth.

Arriving at the planet at about the same time as the probe, the main spacecraft will pass 133,000 miles above Jupiter's cloud tops, then swing into orbit around the planet. Galileo will then begin

two years of observation of Jupiter, its moons and its powerful magnetosphere. Galileo is a cooperative U.S./Germany project managed for NASA by JPL, which designed and built the main spacecraft. Ames Research Center has management responsibility for the probe, which was built by Hughes Aircraft.

Ulysses and Galileo were two of six NASA deep space projects active during 1995. The others were three veteran spacecraft — Pioneer 10, Voyagers 1 and 2 — operating billions of miles from Earth and conducting scientifically important research, looking for the boundary between the solar system and interstellar space.

Another major planetary exploration project advanced in 1995 as static testing began on components of the Cassini spacecraft. Cassini is a joint project of NASA, ESA

and the Italian Space Agency, managed for NASA by JPL. To be launched in 1997, the Cassini spacecraft, consisting of an orbiter and a probe, will conduct an extensive survey of Saturn and its principal moon Titan, beginning in 2004.

The large (750-pound) ESA-built instrumented probe — named Huygens — will descend into the atmosphere of Titan, which is believed to be chemically similar to the atmosphere of early Earth and is therefore of immense scientific interest.

The 24-ton Cassini main spacecraft will conduct a four-year orbital survey of Saturn, during which time it will orbit the planet about 60 times and encounter Titan 30 times; the spacecraft will provide data and imagery on the planet, its moons, rings and magnetosphere.

(Continued)

NASA's Discovery program is an ongoing effort to foster development of low cost spacecraft that will enable frequent solar system exploration missions. The flight program will get under way early in 1996 with the launch of the Near Earth Asteroid Rendezvous (NEAR). The NEAR spacecraft will rendezvous with the largest of the near-Earth asteroids — 433 Eros — and orbit the asteroid for at least a year, beginning in 1999. It will collect a wealth of data on these little-known celestial bodies.

Later in 1996, the second of the Discovery spacecraft, Mars Pathfinder, will be boosted into a trajectory toward the Red Planet. Pathfinder is designed to demonstrate the technology, systems and mission elements involved in landing a series of small surface stations and rovers on Mars. Pathfinder will land on Mars in mid-1997 and dispatch a "minirover," a small, inexpensive vehicle designed to roam the Martian surface and send back to Earth three-dimensional images of the Marscape.

Above is a sequential concept of a Pathfinder landing. The "balloons" are air bags to cushion the landing. In the final sequence, the space-craft has landed and is unfolding its petals to expose its solar panels and release the rover.

In February 1995, NASA added a third Discovery developmental project, the Lunar Prospector. Scheduled for launch in June 1997, Lunar Prospector will map the chemical composition of the lunar surface and the Moon's magnetic and gravitational fields at a level greater than that achieved by any previous lunar mission. The mission will also seek to locate any significant quantities of water that might exist in shadowed craters near the lunar poles, a factor of importance in planning further human exploration of the Moon.

The Lunar Prospector **(right)** is being built by Lockheed Martin Missiles and Space Company and it will be launched by the company's new Lockheed Launch Vehicle. Ames Research Center is responsible for technical support and for one of the spacecraft's instruments.

In 1996, NASA will resume the comprehensive exploration of Mars that began with the Viking orbiters/landers of the late 1970s. The Mars Surveyor program, managed by Jet Propulsion Laboratory (JPL), consists of an initial mission by an orbiter known as the Mars Global Surveyor, followed by a series of orbiter/lander pairs to take advantage of launch opportunities that occur about every two years as Mars comes into alignment with Earth.

Scheduled for launch in November 1996, the Mars Global Surveyor will carry five science instruments to conduct a systematic mapping of the planet from polar orbit. The mission is intended to provide most of the science data that would have been supplied by the Mars Observer spacecraft lost in 1993; the instrument complement is comprised largely of spare parts from the Mars Observer. Compared with previous planetary explorers, Global Surveyor is small at one ton liftoff weight. The spacecraft is being built by Lockheed Martin Astronautics.

In March 1995, NASA selected the second set of spacecraft in the Mars Surveyor program. They are two small complementary spacecraft known as Mars Surveyor 1998 Orbiter/Lander. The label pretty much describes the project: the pair will be launched in the Mars opportunity that spans two months beginning in December 1998. The orbiter will be launched in December 1998 and go into orbit around the planet in September 1999. The lander will be launched in January 1999, enter the Martian atmosphere in October of that year, descend by parachute and land softly with the help of rocket thrusters. The two craft will team in a study of the planet's climate and a search for water in the Martian soil.

Lockheed Martin Astronautics is building both vehicles of the Mars Surveyor 1998 mission. The project exemplifies the trend toward smaller, less expensive NASA spacecraft. Both will be built within a funding level previously allocated for a single mission. The 1996 orbiter will weigh only half as much as the 1996 Mars Global Surveyor; the lander will weigh only half as much as the 1996 Mars Pathfinder, the smallest planetary lander yet constructed. The low weight of the craft will allow them to be launched on a new vehicle, now in development, called Med-Lite, which is roughly half the size of the Delta II launch vehicles that will boost the 1996 Global Surveyor and Pathfinder missions.

(Continued)

ASTRONOMY AND ASTROPHYSICS

On April 24, 1995, NASA's Hubble Space Telescope (HST), widely acknowledged as the finest astronomical instrument ever built, celebrated its fifth anniversary in orbit (the HST is shown **at right** in the payload bay of the Orbiter *Endeavour* as astronauts completed the first Hubble servicing and refurbishment mission in 1993). Just prior to the anniversary, the HST provided astronomers with another surprising major discovery, a new great dark spot in the cloudy northern hemisphere of the planet Neptune.

Months earlier — in June 1994 — Hubble images had disclosed that a large dark spot in Neptune's southern hemisphere, originally discovered by Voyager 2 in 1989, had mysteriously disappeared. The new dark spot found in 1995 is very similar to the feature first mapped by Voyager 2, but located in a different part of the planet. Astronomers say that the Hubble finding shows that Neptune has changed radically since 1989, indicating that Neptune has an extraordinarily dynamic environment in which the planet can look completely different in only a few weeks.

In June 1995 Hubble investigators announced that the

HST had detected a vast disc of perhaps 200 million large comets orbiting beyond the orbit of Neptune. In the same month, the HST confirmed the existence of primordial helium in the universe, a discovery that supports the theory as to the creation of hydrogen and helium in the immediate aftermath of the Big Bang. These discoveries were among hundreds of important findings Hubble has contributed in the brief span of five years. With a resolving power calculated to be 10 times better than any telescope on Earth, the HST has compiled a rate of discovery that is unprecedented for any modern observatory. NASA Associate Administrator for Space Science Wesley T. Huntress Jr. has termed the telescope "truly a national scientific treasure." The Hubble is a cooperative program of NASA and the European Space Agency (ESA). Hubble program and project scientists have selected a

"Top Ten" list of the observatory's leading discoveries since it became operational in 1990:

- The first conclusive evidence of the existence of black holes, which have millions or billions the mass of the Sun.

- Indications that the universe might be much younger than previously thought, a finding acquired by Hubble's accurate measurement of the distance to a remote galaxy, which provided a basis for calculating the expansion rate of the universe.

- The first direct visual evidence that the universe is evolving as predicted by the Big Bang theory.

- The discovery that quasars, very distant and extremely bright objects, are even more mysterious

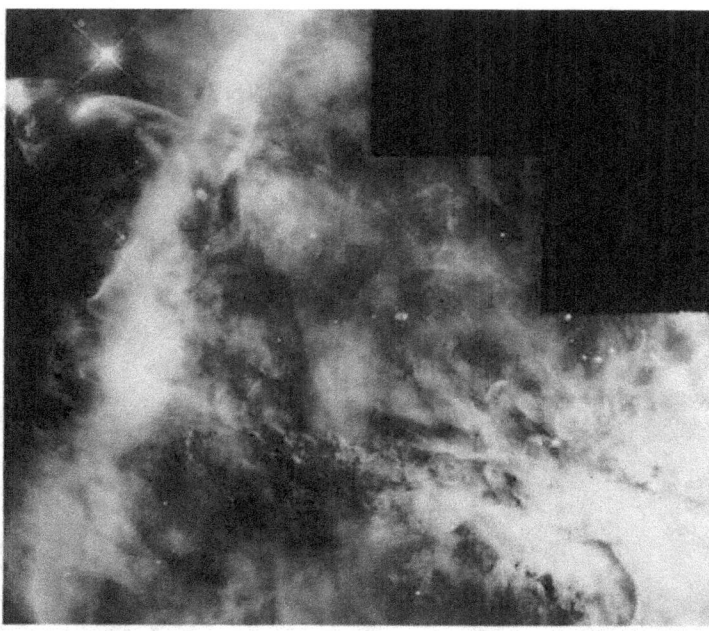

than earlier thought, because many do not dwell in the cores of galaxies but are isolated in space.

- Indications that dark matter in the universe is more exotic than once thought, through a finding that nature does not produce enough of the extremely small Red Dwarf stars that were once considered a leading candidate for the universe's "missing mass."

- Further support of the Big Bang theory by refining estimates of the amount of deuterium in space (deuterium is an element created in the initial cosmic fireball that gave birth to the universe).

- A solution to the mystery of intergalactic clouds of hydrogen, provided by the observation that they are really gigantic halos of galaxies.

- An implication that planets — and presumably life — might be abundant in the universe, based on the discovery of discs of dust that might be embryonic planetary systems around young stars. **Above,** an immense wall of glowing gases forms a colorful backdrop to dozens of newborn stars, many of which have dust discs.

- A contribution of important detail and surprising findings relative to the spectacular 1994 collisions of Comet Shoemaker-Levy with the planet Jupiter.

- The revelation of dynamic weather changes on nearly all the planets, with a clarity once obtainable only by spacecraft fly-bys. Hubble is serving as an interplanetary "weather satellite;" an example of its imagery is shown **at left,** a spring image of Mars' northern hemisphere showing that the carbon dioxide frost around the permanent water-ice cap has sublimated and the cap has receded to a core of solid water-ice several hundred miles across. The image also shows an abundance of wispy white clouds, indicating that the atmosphere was cooler at the time of this image than on visits of space probes in the 1970s.

The Hubble Space Telescope was designed to work in orbit for 15 years. The 1993 refurbishment mission was the first of four planned; a second is scheduled for February 1997 and a third in November 1999.

The largest scientific payload ever orbited, the 12 ½-ton, 43-foot HST was developed by Lockheed Missiles and Space Company and Perkin-Elmer Corporation (now Hughes Danbury Optical Systems) under the management of Marshall Space Flight Center. When the telescope became operational in 1990, Goddard Space Flight Center took over responsibility for controlling the observatory and processing the imagery and data. The center for collection and distribution of the data is the Space Telescope Science Institute in Baltimore, Maryland, operated for NASA by the 17-member Association of Universities for Research in Astronomy.

The Hubble Space Telescope (see page 26) is the first of a series of spacecraft known as the Great Observatories, each of which is designed to operate in a certain range of the electromagnetic spectrum, including the full ultraviolet, infrared, x-ray and gamma ray ranges. This capability is important to astronomical science because each band of the spectrum offers a different set of clues to the origin and evolution of the universe.

The second of the Great Observatories is the Compton Gamma Ray Observatory **(above).** Launched in 1991 and managed by Goddard Space Flight Center, the Compton Observatory is investigating the most energetic of all forms of radiation. Like Hubble, Compton has been regularly making new discoveries, such as the 1994 finding described as "one of the most spectacular astrophysical discoveries of the decade: evidence that gamma ray bursts occur in the far reaches of the universe, bear an imprint of the universe's expansion, and occur so far away that they show relative time dilation." This finding provided additional evidence that gamma ray bursts are not limited to the Milky Way Galaxy.

The third of the Great Observatories is the Advanced X-ray Astrophysics Facility (AXAF), in development and scheduled for launch in September 1998. AXAF **(top right)** passed a major developmental milestone in the spring of 1995: completion of the polishing and measurement of the observatory's eight mirrors, one of the project's toughest technical challenges. The job was completed four months ahead of schedule by the optics contractor, Hughes Danbury (Connecticut) Optical Systems. The spacecraft principal contractor is TRW Inc. The AXAF team also

includes the Smithsonian Astrophysical Observatory and project manager Marshall Space Flight Center.

AXAF is expected to significantly improve scientific understanding of some of the most energetic and violent processes in the universe. It will provide images and spectrograms that will yield information on the temperature and composition of such objects as neutron stars, black holes, debris from supernova explosions, quasars, the centers of active galaxies, and hot gas in individual galaxies and clusters of galaxies.

The contemplated fourth Great Observatory is the Space Infrared Telescope Facility (SIRTF), which is intended to conduct advanced investigations of prime interest targets identified by earlier infrared observatories. Managed by Jet Propulsion Laboratory, SIRTF is in study status in anticipation of hardware development beginning in 1998 and

orbital service in 2002.

Another infrared observatory in development is the Stratospheric Observatory For Infrared Astronomy (SOFIA), a joint project of NASA and the German space agency DARA. SOFIA **(above)** is an airborne rather than an orbital observatory. It consists of a 2.5 meter telescope mounted in a specially modified Boeing 747SP transport; it is planned for initial operation in late 2000.

SOFIA will replace NASA's aging Kuiper Airborne Observatory, a modified telescope-equipped C-141 transport that has been in service since

1974. SOFIA's telescope will be three times the diameter and about 10 times more sensitive than the Kuiper system. The new observatory will enable significant advances in the study of the solar system, star and planet formation, the makeup of the interstellar medium, galactic structure and evolution.

MISSION TO PLANET EARTH

Planet Earth is changing and the consequences of change could have serious implications for human health, the world's food and energy supplies, and the global economy. There is need for greater understanding of the complex mechanisms that control Earth's behavior. That need is being met by a coordinated international research program designed to reduce the uncertainties of global change.

NASA's part of the program is the Mission To Planet Earth (MTPE), a space science effort of such importance that it has been elevated to status as a separate program office. MTPE, which involves extensive cooperation and contributions from other U.S. federal agencies and other nations, employs satellites and other tools to generate data about Earth; it is intended to expand understanding of how natural processes affect humankind, and how humans may be effecting those processes. MTPE studies are expected to yield improved weather forecasts, tools for managing forests and agriculture, information for fishing fleets and coastal planners, and — eventually — an ability to predict how the climate will change in the future.

Phase I of the MTPE program has been under way since September 1991. Among the major contributors is the U.S./France satellite TOPEX/Poseidon which addresses how ocean circulation influences global climate by mapping the circulation patterns of the world's oceans over several years.

TOPEX/Poseidon is an altimeter-type satellite; the illustration **above** shows how such satellites, working with tide-gauge networks and land-based and orbital geodetic location systems, can enable accurate, long-term, continuous monitoring of climate and sea level.

TOPEX/Poseidon's radar altimeter has demonstrated a capability for very precise measurement of sea surface heights and satellite data indicates that sea levels increased during 1994-95. Although the data is not sufficient to establish a trend, sustained increases in sea levels could worsen the effects of hurricanes and other storms, threaten coastal regions, and possibly indicate global warming.

In a related discovery, TOPEX/Poseidon provided data that enabled scientists to track disturbances caused by the lingering effect of the El Nino climate event of 1991-93. Although scientists don't know why El Nino occurs, records show that it has been happening for hundreds of years, profoundly affecting weather patterns and causing floods and droughts in various parts of the world. Over the past decade, the pattern of El Nino occurrences has increased dramatically; some researchers believe the 1993 Mississippi and 1995 California floods were caused by El Nino.

Another MTPE tool is the Shuttle-based Space Radar Laboratory (SRL), which flew two 1994 missions that enabled an international team of scientists to observe shifting boundaries between temperate and boreal (northern) forests. The results of these observations permit study of how such changes might affect climatic change; they are also being used for mapmaking, study and interpretation. NASA and Canadian scientists also used SRL data, along with aircraft and ground data, for the Boreal Ecosystem/Atmosphere Study

of northern forests, a large scale investigation into how the forests and the atmosphere interchange energy, heat, water, carbon dioxide and other trace gases.

Data from the Upper Atmosphere Research Satellite, which is investigating the role of the upper atmosphere in climate and climatic change, provided conclusive evidence that human-produced chemicals, including chlorofluorocarbons, are breaking down in the stratosphere and causing depletion of ozone over the Antarctic.

Another satellite-based instrument, the Total Ozone Mapping Spectrometer (TOMS) aboard the Russian Meteor-3 satellite, reported that the 1994 Antarctic "ozone hole" was the largest on record. However, measurements by NASA and other agencies indicated that control measures to protect the ozone layer were beginning to work and that the atmospheric concentration of damaging chemicals was stabilizing.

Phase II of MTPE will begin in 1998 with the launch of the first Earth Observing System (EOS) spacecraft, EOS AM-1 **(above).** EOS is the first integrated satellite/surface

research system for observing how the Earth's parts (air, water, land and life) interact with each other. Studying these interactions is the critical next step in understanding the complexities of the global climate. EOS will contribute to such understanding by observing Earth in 24 measurement areas over 15 years, employing instruments on satellites of varying size and configuration. Each spacecraft will focus on a different aspect of global climate change and the interrelation among Earth's environmental components. EOS data, along with data from airborne and ground-based systems, will enable scientists to model Earth as a global system, thus fostering fundamental advances in weather and climate forecasting, land use planning, disaster relief and, ultimately, providing the answers to scientists' many important questions about long-term climate changes.

NASA's aeronautical research program is providing a technology base for tomorrow's safer, more efficient, more competitive aircraft

In 1915, the Congress established the National Advisory Committee for Aeronautics (NACA) as an independent agency of the federal government "to supervise and direct the scientific study of the problems of flight with a view to their practical solution, and to direct and conduct research and experiments in aeronautics."

For 43 years, until it was absorbed by NASA, NACA was extraordinarily successful in carrying out the Congressional mandate, contributing a lengthy list of technological advances and innovations that helped thrust the U.S. aircraft and airline industries into positions of leadership in world aviation. Today, 80 years after NACA's creation, NASA's aeronautics program follows in NACA's grand tradition, providing a continuing flow of advanced technology over a very broad spectrum of aviation-related disciplines. This program represents an investment in the future of U.S. aircraft manufacturing capabilities and air transportation safety and efficiency, areas that rank among the leading influences on the national economy, security and public convenience.

The competitiveness of the aircraft industry is a matter of particular importance. Consensus puts the market for commercial jetliners over the next 20 years at more than one trillion dollars; if that potential is realized, it could mean worldwide production of airliners at an annual rate roughly double that of the record-setting years of the 1980s. But American plane builders have lost significant market share to foreign competitors in recent years and the competition continues to intensify.

What is needed to preserve American competitiveness — and the aircraft sales, jobs and tax revenues that go with it — is a continuing infusion of advanced technology. It is the government's role — in this instance NASA's — to invest in the higher risk, longer term technologies that otherwise might go undeveloped because the payoffs are too uncertain and too distant to justify corporate outlays.

U.S. aeronautics and air transportation are challenged on other fronts. There is need for continual improvement in aircraft and airways safety, a corollary need for improving the national airspace capacity, and a need for enhancing the environmental characteristics of the airplane. Here again, advanced technology is the answer.

Finally, NASA must maintain a national research infrastructure for government and industry, including wind tunnels, simulators, computational facilities, and flight research platforms. This infrastructure demands continuing modernization.

NASA, therefore, is engaged in a broad aeronautical research and technology program that involves projected outlays of almost $5 billion over the fiscal years 1996-2000. That investment is expected to pay important dividends to the nation. It will help eliminate the $3.5 billion a year wasted due to airport delays; posture U.S. industry for leadership in the 21st century supersonic transport market estimated at $200 billion; and strengthen American industry's competitiveness in the trillion dollar subsonic market.

Symbolic of NASA's trailblazing aeronautics program is an investigation of a futuristic "waverider" aircraft that would fly at 3,000 miles per hour, riding on its own shockwave as surfboards ride on ocean waves. In the photo, a Langley Research Center engineer is readying a waverider model for wind tunnel test.

NASA's aeronautics goals, as detailed in 1995 Congressional testimony, are to

- Develop high payoff technologies for a new generation of economic, environmentally acceptable U.S. subsonic aircraft and a safe, highly productive air transportation system;

- Build the technology base for an economically viable and environmentally friendly high speed civil transport;

- Explore technology options for new capabilities in high performance aircraft;

- Develop and demonstrate technologies for airbreathing hypersonic flight;

- Develop advanced concepts, physical understanding, and theoretical/ experimental/ computational tools to enable advanced aerospace systems; and

- Develop, maintain and operate critical national facilities for aeronautical research in support of industry and technology-developing government agencies.

NASA pursues these objectives through in-house research and through cooperative endeavors with academia, industry and other government agencies. NASA's principal research installations are Ames Research Center, Moffett Field, California; Dryden Flight Research Center, Edwards, California; Langley Research Center, Hampton, Virginia; and Lewis Research Center, Cleveland, Ohio. Examples of their activities are described on the following pages.

HIGH SPEED RESEARCH

Aircraft manufacturers of several nations are developing technology to position themselves for the next plateau of international aviation competition: the long range, economical, environmentally acceptable second generation supersonic passenger transport, which could be flying before 2010.

Market experts predict that a projected quadrupling of traffic to and from the nations of the Pacific Rim, together with more moderate increases in demand for long range passenger transportation in other areas of the world, will create a need for some 500 next generation supersonic transports worth an estimated $200 billion and 140,000 jobs.

Capturing a major share of that market is vitally important to a U.S. aerospace industry that is transitioning from a traditionally defense-oriented product line to a commercially-driven manufacturing activity. To help boost the industry's competitiveness, NASA is conducting a High Speed Research (HSR) program that addresses the highest priority, highest risk technologies needed for a High Speed Civil Transport (HSCT). The HSR program is intended to demonstrate the technical feasibility of the vehicle; the decision to proceed with full scale development will be up to industry and will hinge primarily on market, financial and other business considerations.

The 21st century HSCT will probably be only moderately faster than the Anglo-French Concorde now in airline service, because studies indicate that the most promising speed — considering propulsion, structural and aerodynamic requirements — is Mach 2.4, or about 1600 miles per hour. However, it will most likely have almost double the Concorde's range and will carry three times as many passengers. Like the Concorde, the HSCT will fly supersonically only over water, but research suggests that it can be economically viable in subsonic flight over land.

The HSR program is being conducted as a national team effort with shared government/industry funding and responsibilities. The team includes NASA's Langley, Lewis and Ames Research Centers and Dryden Flight Research Center; engine manufacturers GE Aircraft Engines and Pratt & Whitney division of United Technologies; airframe manufacturers The Boeing Company and McDonnell Douglas Corporation; other manufacturers, materials suppliers and academic institutions.

The team has established a baseline design concept, known as Reference H, which serves as a common configuration for HSR investigations. Shown being readied for a wind tunnel run **at left** is a 19-foot model of the Reference H design; **at right,** a smaller model is being flown in Langley Research Center's National Transonic Facility. A full-size airplane of this type would be capable of accommodating 300 passengers and flying route segments as long as 5,000 nautical miles.

Phase I of the HSR program, which focuses on environmental challenges — engine emission effects on the atmosphere, airport noise and sonic boom — got under way in 1990 and will continue through 1995.

A most critical technology need is an advanced combustion concept to reduce the nitrogen oxide (NOx) emissions expelled in jetliner exhaust, because NOx can cause an ozone-depleting chemical reaction in the atmosphere. The HSR team has established a NOx index goal that is some 90 percent below current emission levels; if that goal can be attained, the stratospheric effect of emissions by a whole fleet of HSCTs would be almost negligible.

The goal for aircraft noise levels is to make the HSCT as quiet as the newest subsonic transports; that goal, though demanding, appears achievable. Similarly, the team has identified promising approaches toward alleviating the sonic boom; current research seeks to determine how much the boom can be softened without significantly compromising aircraft performance. Generally speaking, although Phase I is still ongoing, NASA officials say there is "growing confidence" that environmental concerns

can be satisfied.

Phase II of the HSR program, initiated in 1994, is running concurrently with Phase I; it focuses on the technology advances needed for economic viability, principally weight reductions in every aspect of the baseline configuration because weight affects not only the aircraft's performance but its acquisition cost, operating costs, and environmental compatibility.

One target area is airframe materials and structures; the HSR team is developing, analyzing and verifying the technology for trimming the baseline airframe by 30-40 percent. In aerodynamic research, a major goal is to minimize drag to enable a substantial increase in HSCT range; Phase II incudes computational and wind tunnel analyses of the baseline HSCT and alternative designs over the entire speed range. Propulsion research looks for environment related and general efficiency improvements in critical engine components, such as inlet systems,

combustors and nozzles. Other research involves ground and flight simulations aimed at development of advanced control systems and flight deck — or technology for optimal supersonic flying quality.

The HSR program will move beyond laboratory investigations into the actual supersonic flight realm through a NASA agreement with the Russian Tupolev design bureau. Under this agreement, Tupolev will modify a TU-144 supersonic transport for use as a flying laboratory gathering aerodynamic structural and environmental data. First flight is targeted for early 1996.

Overall, the HSR program seeks technology advances that would make a 2005 HSCT almost 50 percent lighter than the baseline HSCT established in 1990 and enable attainment of the economic and environmental goals.

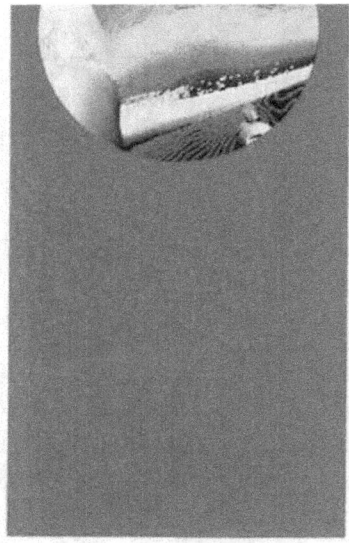

ADVANCED SUBSONIC TECHNOLOGY

Although the High Speed Civil Transport of the 21st century (see page 34) may dominate airline service on long overwater routes, the subsonic jetliner will continue to handle most overland travel. The subsonic aircraft market — estimated at $1 trillion over the next two decades — will constitute the bulk of the total market.

U.S. manufacturers hold a 67 percent share of that market, but their share has eroded considerably since 1985 and the principal competitor, the European Airbus consortium, expects to capture half the total market by 2000. It would be a major blow to the U.S. economy should that happen.

The U.S. aviation industry faces other challenges. U.S. airlines estimate that capacity limitations of the National Airspace System cost them $3.5 billion a year in delayed flights. Having experienced heavy financial losses in recent years, the airlines are flying their aircraft longer, a situation that demands new technology to ensure safety over a jetliner's longer life span that is in many cases 30 years. There is further demand for new technology to meet increasingly stringent environmental regulations in the U.S. and abroad.

Therefore, NASA, other government agencies and U.S. plane builders are engaged in an Advanced Subsonic Technology (AST) program that has two strategic thrusts: developing high leverage technologies that will assure the future competitiveness of U.S. civil transports, and finding new ways to enhance the safety, productivity and environmental acceptance of the national air transportation system. AST research involves the whole spectrum of investigational facilities, including computational facilities, laboratories, wind tunnels, propulsion test beds, and flight test. **The photo at left** exemplifies the wind tunnel testing being accomplished, in this case a wing and a nacelle of the new Boeing 777 twinjet. **At right** is Langley Research Center's Transport Systems Research Vehicle, a Boeing 757 jetliner modified for flight test of advanced navigation and landing systems, flight controls, weather aids, systems for improving terminal area productivity and other airborne equipment.

Among the areas being addressed under the AST program are:

- Aging Aircraft. An area of focus is development of advanced non-destructive evaluation techniques to detect disbonds, fatigue cracks and corrosion, and to enhance aircraft inspection techniques by developing analytical tools for quantitative evaluation of inspection findings and for predicting the residual safe structural life of an aging airplane.

- Environmental Assessment. The impact of aviation on the atmosphere is becoming more and more a matter of concern in developed nations and jet

liner cruise emission standards are being considered. To assure that the interests of U.S. aviation are fairly represented in international regulatory considerations, NASA is conducting a comprehensive scientific assessment of emission effects on the upper troposphere and the lower stratosphere. The goal is to develop technology, by 2000, that will provide a U.S. capability for assessing changes in atmospheric ozone caused by aircraft emissions.

- Noise Reduction. NASA's aim is to develop technology that will enable an airplane noise reduction of 10 decibels below the mean level of 1992 in order to help achieve compliance with increasingly stringent noise standards.

- Terminal Area Productivity. NASA seeks to develop technology for achieving the instrument weather capacity of the National Airspace System to the point where it is equivalent to clear-weather capacity. One goal is a 12-15 percent increase in capacity over that of current non-visual operations on a single runway; another is to demonstrate the feasibility of increasing capacity on parallel runways by reducing the spacing between aircraft.

- Propulsion. AST work involves development of concepts for reducing ozone-depleting nitrogen oxide emissions; and improving fuel consumption for future commercial aircraft engines.

- Composite Materials. Composites offer potential for reducing aircraft weight, hence cutting costs, but composites have so far been limited to secondary aircraft structures, such as wing leading edges, flaps, ailerons, fairings and access doors. Researchers feel that the technology can be extended to primary structures, which could produce significant cost benefits and competitive advantage. NASA's goal is development and ground test of a full-span composite wing and center section by 2000.

- Integrated Wing Design. This area involves designing wing components in an integrated manner to reduce cost and design time while improving aircraft performance, NASA seeks more efficient high lift devices for low speed takeoffs and approaches with minimal noise impact on the community, together with drag-reducing laminar control of the airflow over the wing.

- Fly-by-Light/Power-by-Wire. Fly-by-light means replacing the current fly-by-wire control systems with lightweight, highly reliable fiber optical systems that eliminate the concerns of electromagnetic interference associated with fly-by-wire. Power-by-wire involves replacing hydraulics with an electrical power actuation system.

- Civil Tiltrotor. The short-haul tiltrotor transport offers promise as a means of transportation to and from city centers and an attendant easing of congestion at outlying hub airports. The AST program addresses technologies that will overcome the "inhibitors" to acceptance of this new types of transport. Target areas include reduction of rotor noise, procedures for steep descents, safe flight with one engine inoperative, and all-weather operations.

HIGH PERFORMANCE COMPUTING

For many years, aeronautical researchers have used computer design techniques to create mathematical models of aircraft and "fly" them by computer simulation, thus allowing study of many different configurations before settling on a final design. Called computational simulation, this technique has expanded enormously, not only in aeronautics but in many other areas of science and engineering.

High performance computing is essential to virtually all of NASA's major programs; it supports all phases of NASA activity from formulation of hypotheses to conceptual design to analysis to mission operations. In aeronautics it is vital not only to NASA's own technology, but to NASA's mission of supporting and enhancing the competitiveness of the U.S. civil aircraft manufacturing industry. In Earth sciences and space sciences, high performance computing has enabled NASA to make significant strides in weather/climate research, galactic evolution studies and solid Earth modeling.

Although NASA has achieved a supercomputing capability of several billion FLOPS (floating points per second), even that tremendous capability will not be adequate for the computational demands of tomorrow, because complex future flight vehicles will be more difficult to simulate. Similarly, existing computer systems are not adequate to the needs of astrophysical investigation and modeling Earth's atmosphere and oceans to determine the extent, causes and consequences of global climate change.

To help bring about the advanced supercomputing capability needed, NASA is conducting a High Performance Computing and Communications (HPCC) program as part of a broader federal HPCC effort intended to support a wide range of economic and social goals, including HPCC networks and advanced software for the National Information Infrastructure. The NASA HPCC program is conducted under the guidance of the agency's Office of Aeronautics; Ames Research Center is the lead center, assisted by Langley Research Center and Lewis Research Center.

The immediate aim of the NASA HPCC program is achievement of sustained computational speeds of *one trillion* FLOPS. NASA, in cooperation with an industry consortium headed by IBM Corporation, is developing applications software and algorithms for scalable parallel computing systems that will enable attainment of the goal by 1997.

The technologies being developed will be applied to designing and simulating advanced aerospace vehicles, including commercial transports; increasing scientists' abilities to model Earth's climate and forecast global environmental trends; advancing astronomy and astrophysics research; and improving the data-gathering capabilities of future spacecraft. Additionally, the NASA HPCC effort seeks to apply computing technologies to such non-aerospace areas as elementary and secondary education; experiments in packet video over the Internet; digital libraries; and applications for remote sensing.

The aeronautics segment of the HPCC program is the Computational AeroSciences component, centered at Ames Research Center. There NASA and the industry consortium are installing additional supercomputers for cooperative testing. The main testbed is Ames' Numerical Aeronautics Simulation (NAS) facility, a national supercomputer complex linked to more than 1,200 industry/university/government scientists by the national computer network AEROnet.

The expanded system at Ames Research Center will

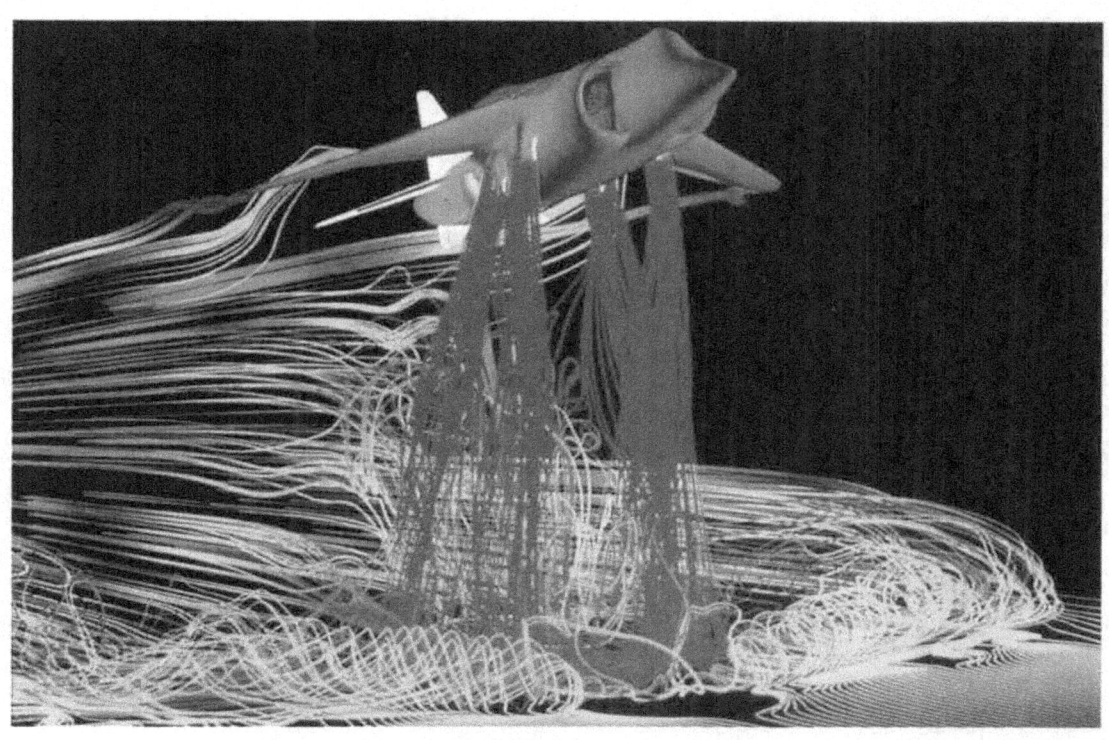

greatly augment other scalable parallel computers to provide research platforms for systems software, virtual wind tunnels, critical disciplines research, and other areas of benefit to government researchers and aerospace manufacturers. At a time when national budget constraints and the interests of U.S. competitiveness dictate sharp focus on cost-reduction measures, the NASA HPCC program offers high potential for economy through faster, more efficient aircraft and spacecraft design.

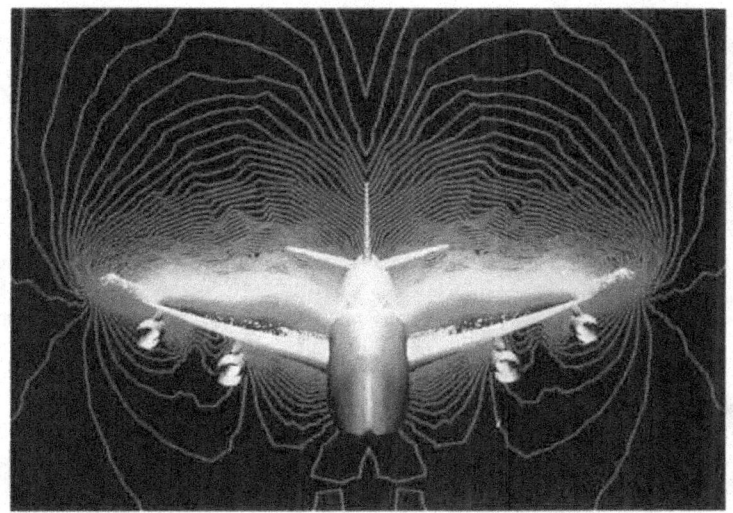

FLIGHT RESEARCH

Shown **below** is the X-31 Enhanced Fighter Maneuverability demonstrator, which in 1995 concluded a multiyear flight test program that accomplished what was termed a "landmark contribution" to military aeronautics. Built by an international team of Rockwell International and the German Deutsche Aerospace, the X-31 made hundreds of flights from 1990 to 1995 demonstrating the potential of thrust vectoring (maneuvering by directing the exhaust flow) and an advanced flight control system for close-in air combat at very high angles of attack (the angle between an air-plane's wing and the air through which it is flying).

The tests were conducted at Dryden Flight Research Center by an International Test Organization (ITO), a component of the NATO Cooperative Research and Development Program, under the management of the U.S. Advanced Research Projects Agency (ARPA). In addition to ARPA and NASA, ITO included representation from the German Ministry of Defense, the U.S. Air Force and Navy, and the co-manufacturers.

The X-31 tests attained all program goals and compiled an impressive array of technically significant "firsts." It demonstrated exceptional agility at extremely high angles of attack, well beyond the normal aerodynamic stall limit; the combat value of its advanced technologies by winning 117 of 128 simulated combat engagements against two of the nation's highest performing military aircraft; and the effectiveness of thrust vectoring for stability and control at supersonic speed, including a successful simulation of controlled flight without a vertical tail.

The X-31 program is representative of the type of flight research conducted by NASA to explore new technologies and new flight regimes. NASA conducts such programs independently or in cooperation with U.S. industry and the Department of Defense, sometimes in cooperation with international development teams.

Another example is the research work performed by NASA's SR-71 **(lower right),** a one-time reconnaissance plane converted to duty as a NASA research platform. In 1994-95, the SR-71 was used on three major test programs: the Optical Air Data Experiment, involving tests of an advanced laser-based angle of attack/airspeed/sideslip measurement system; a Handling Qualities Experiment, in which the SR-71 served as a stand-in for the reference model High Speed Civil Transport (HSCT); and a sonic boom test program designed to measure the effects of boom pressure waves under various meteorological conditions at various altitudes.

The SR-71 flights were conducted for Langley Research Center's High Speed (HSR) program (see page 34) with the goal of building a database to help engineers predict the shape and intensity level of sonic booms generated by future high speed aircraft, such as the HSCT. The SR-71 was employed as the boom generator, while an instrumented F-16XL aircraft probed the boom's shock wave from a position several hundred feet from the SR-71. Data was also collected by two other aircraft at lower altitudes and by sensors on

the ground. Data on the characteristics and intensity of the booms may help designers shape future supersonic aircraft to soften the boom effect.

While one F-16XL was engaged in the boom tests, another — the F-16XL Supersonic Laminar Flow Control (SLFC) aircraft **(right)** — was participating in a different aspect of Langley's HSR program. The SLFC aircraft is fitted with a "suction glove" on its left wing. The glove, made primarily of titanium, has millions of microscopic laser-drilled holes. The objective is to pull turbulent air through the holes by means of a suction system, thereby creating a laminar (smooth) flow over the wing. Laminar flow lowers air drag and permits faster speed, longer range or reduced fuel consumption. The 1995 flights represented a continuation of a program in flight test status since

1992, but the new series involved evaluation of a new laminar flow control system, one that more closely resembles the type of apparatus that would be installed on a HSCT.

A veteran research aircraft, NASA's F-18 HARV (High Alpha Research Vehi-

cle) embarked on a new phase of an investigation that has been in flight test since 1987. High alpha refers to the high angle of attack, in which the nose is angled up as much as 70 degrees while the aircraft flies horizontally. This is called a post-stall maneuver, because the aircraft would stall and spin without the control assistance supplied by its thrust vectoring system. After several series of successful tests with the thrust vectoring system alone, the HARV was fitted with a nose-mounted strake system. The new series of flights is evaluating how the pilot-actuated strakes improve control at high alpha when coupled with the thrust vectoring system already on the HARV.

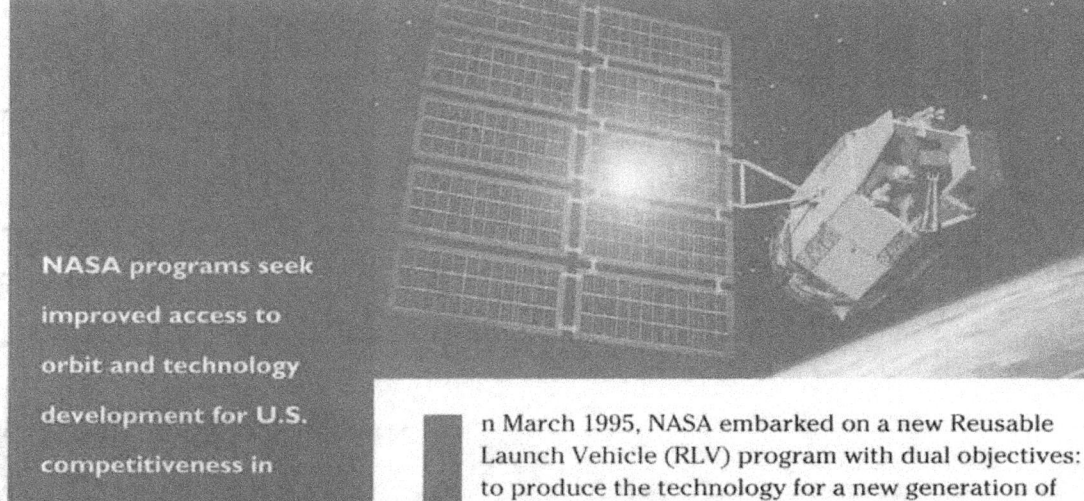

NASA programs seek improved access to orbit and technology development for U.S. competitiveness in space systems

n March 1995, NASA embarked on a new Reusable Launch Vehicle (RLV) program with dual objectives: to produce the technology for a new generation of space boosters capable of delivering payloads to orbit at significantly lower price, and to provide a technology base for development of advanced launch systems that will make U.S. aerospace manufacturers more competitive in the global commercial market.

The RLV program also represents a new way of doing business at NASA. The vehicles are being developed on a government/industry partnership basis wherein NASA and the participating manufacturing companies share development costs.

NASA signed four agreements with industry firms for technology development efforts that could lead to commercially available U.S. reusable launch systems early in the 21st century. They would be the first completely new launch vehicles developed in the U.S. since the introduction of the Space Shuttle in 1981.

The initial development is an RLV designated X-34, being developed by American Space Lines, a joint venture of Orbital Sciences Corporation (OSC), Dulles, Virginia and Rockwell International's Space Systems Division, Downey, California (Rockwell built the Space Shuttle Orbiters). Designed to deliver relatively small payloads (1,000 pounds), the X-34 is a reusable winged vehicle launched from an aircraft. After a rocket burn of approximately three minutes, the X-34 releases a payload vehicle that flies independently into orbit and is not recovered. The main X-34 vehicle returns to Earth for a wheeled landing on a conventional runway.

Being developed by Orbital Sciences Corporation and Rockwell International, the X-34 small reusable booster will be air-launched from a 747 transport; it will deliver a payload to orbit, then reenter the atmosphere and land on a runway.

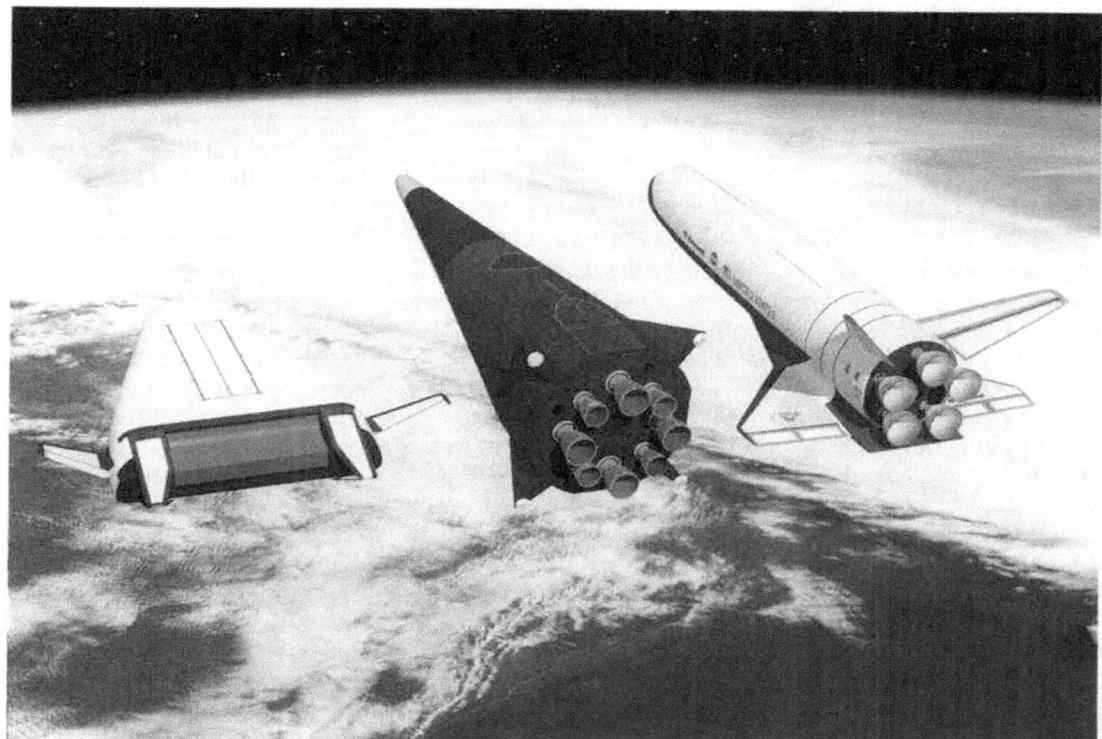

Under an accelerated development schedule, the OSC/Rockwell team expects initial launch of the X-34 in 1998. The craft will serve as a test bed for demonstrating RLV technology and as an operations pathfinder for a larger RLV known as the X-33.

NASA signed cooperative agreements with three companies for X-33 development: Rockwell Space Systems; Lockheed Martin Advanced Development Company, Palmdale, California; and McDonnell Douglas Aerospace, Huntington Beach, California. The agreements call for joint NASA/company work during a 15-month Phase I concept definition and design effort, during which each contractor will develop its own vehicle design, operations plan and business investment strategy.

The results of Phase I will provide the basis for a White House decision in 1996 as to whether to proceed with Phase II, which involves design, construction and flight testing of an X-33 vehicle. If the decision is affirmative, NASA will select one or more of the participating companies as its industry partner(s) for Phase II. Phase II would continue through the end of the decade, at which time government and industry would jointly decide whether to proceed to development of an operational, next-generation RLV.

The Reusable Launch Vehicle Technology Program is managed by the Office of Space Access and Technology in NASA's Washington, D.C. headquarters. Marshall Space Flight Center is host center for both the X-34 and X-33 programs. Government laboratories will work with each of the design teams to apply technology developed in the labs to the new launch systems.

ADVANCED COMMUNICATIONS TECHNOLOGY

Existing commercial space communications systems evolved from high risk technology developed and tested in orbit by NASA in the 1960s and 1970s. The technology base thus provided enabled the U.S. telecommunications industry to lead the way in developing, manufacturing and operating commercial communications satellites,

with a resultant benefit to the U.S. economy.

Today, as information technologies grow explosively, there is need for a new infusion of space communications technology to meet

requirements for new types of space communications and data transfer, and to allow more efficient use of orbit and spectrum resources. Responding to that need, NASA is demonstrating new technologies with the innovative Advanced Communications Technology Satellite (ACTS) shown in **below.**

A large satellite weighing more than 15 tons and stretching 47 feet across the width of its solar array, ACTS operates in the Ka-band, a little used portion of the radio

frequency spectrum; its on-board transponders allow very high data transmission rates. ACTS also features significant capacity gain through use of multiple hopping beam antenna systems,

which generate message-carrying spot beams, each focused on a narrow Earth region as opposed to the broad beams generated by earlier satellites. Other advanced technologies include on-board digital processing and storage/switching of satellite signals.

ACTS completed its second year in orbit on September 12, 1995. During that time, more than 80 organizations logged an estimated 16,000 hours performing field trials, such as testing the spot beams that enable communications with small, low-cost Earth stations, or ACTS' on-demand digital services that allow users to pay for bandwidth only when needed. These and other ACTS technologies combine to offer potential for significant cost reductions that would make practicable such new services as remote medical imaging diagnostics, global personal communications, real-time TV transmission to airliners, direct transmission of image data to battlefield commanders, interconnection of distantly-located supercomputers, and high-speed off ramps for the National Information Infrastructure.

ACTS was built for NASA by Martin Marietta Astro Space, East Windsor, New Jersey, under the management of Lewis Research Center.

TELEROBOTICS EXPERIMENT

For a decade or more, NASA has been conducting a program designed to facilitate future space exploration by employment of telerobotic systems. A major advancement is development of a new vehicle — known as Ranger — designed as a low cost means of moving telerobotics experiments beyond the laboratory to actual space flight experience.

Shown **at right**, Ranger is a dual arm, free flying telerobotics experiment for on-orbit validation of many of the technologies developed in the NASA program. The Ranger Telerobotic Flight Experiment is coordinated by the University of Maryland Space Systems Laboratory, in cooperation with Jet Propulsion Laboratory, Lewis Research Center and the Michigan Space Automation and Robotics Center. Ranger will carry experiments from every NASA center involved in the Telerobotics Research Program, plus experiments from industry and other universities.

Two Ranger vehicles are being built. The first, designed to operate under water, will undergo extensive testing at the Neutral Buoyancy Research Facility on the University of Maryland campus to get basic data on its operation and capabilities. The second, as nearly identical to the first as possible, is a flight vehicle, scheduled for launch in late 1996.

The project will correlate neutral buoyancy robotic simulation by nearly identical underwater and flight units performing identical tasks in both environments; that will increase understanding of the capabilities and limitations of existing techniques for simulating the space environment on Earth.

On orbit, Ranger will demonstrate a variety of space operations tasks, from the relatively simple installation of Orbital Replacement Unit modules to complex satellite servicing/refueling tasks that have thus far only been performed by astronauts in extravehicular gear. Utilizing telepresence ground-based control, coordinated manipulation operations, automated rendezvous and docking technology and a hybrid propulsion system, Ranger will conduct simulated satellite servicing exercises to characterize the operational capabilities of free flying robotic systems.

Ranger represents the first of a new class of low-cost expendable robots designed for research and servicing in areas beyond the reach of the Space Shuttle. The Ranger vehicles will incorporate design considerations for advancing technical education in the U.S. by encouraging direct student involvement in space research.

ORBITAL VACUUM FACILITY

In the increasingly competitive global aerospace market of the 1990s, American manufacturers are intensifying their research on electronic products and components to gain economic advantage. A key area of this effort is research on advanced semiconducting materials. In the interest of U.S. competitiveness, NASA is providing industry a technological boost with a materials research program of exceptional potential.

The majority of electronics components in use today are made of the semiconductor silicon. However, there are many other semiconductors with much higher predicted performance than silicon, for example, gallium arsenide and aluminum gallium arsenide. Devices made from these materials could provide higher speeds and superior optical response, consume less power, and operate at higher temperatures, thereby offering important advances in superspeed computers, lasers, communications and infrared systems, and many other high technology microelectronics applications.

The problem is that current materials available do not have the requisite structure and high enough quality to attain predicted perfor-

mance levels. To improve their quality, materials can be grown as crystalline thin films in a vacuum chamber — but this technique, known as epitaxy, is limited by vacuum conditions in Earth-based chambers. A key to improving semiconductor materials, therefore, is to improve the vacuum environment for thin film growth.

That is the purpose of the Wake Shield Facility (WSF) program being conducted by the Space Vacuum Epitaxy Center (SVEC) of the University of Houston, one of NASA's Centers for the Commercial Development of Space, in cooperation with a consortium of industrial partners led by Space Industries, Inc. (SII), Houston, Texas. The aim of the program is to demonstrate that low Earth orbit (LEO) offers an "ultravacuum" for growing electronics materials of significantly higher quality than can be produced on Earth, and that these materials can

be processed *in situ,* pointing the way toward future orbital manufacturing facilities producing increasingly sophisticated materials.

The WSF is an orbital laboratory that makes space epitaxial research possible. Carried in the payload bay of the Space Shuttle Orbiter, it is a two-part system consisting of a support unit known as the Shuttle Cross Bay Carrier and a free-flying spacecraft. Released from the Orbiter, the WSF Free Flyer becomes a separate orbiting vehicle **(below)**, pushing the few atoms present in the moderate LEO vacuum out of the way and creating a wake in which there are virtually no atoms. The wake ultravacuum is 1,000 to 10,000 times better than those in the best Earth vacuum chambers. Additionally, the wake region exhibits a nearly infinite pumping speed, removing the excess materials that contaminate Earth chambers.

Built by SII and the SVEC

team, the 8,100 pound WSF package takes up one quarter of the Orbiter's payload bay, mounted horizontally with the Carrier on the bottom and the 12-foot-diameter disc-shaped Free Flyer atop it, connected by a latch system. The 4,350-pound Free Flyer has 60 kilowatt hours of energy, stored in silver-zinc batteries, to power the thin film growing cells, the substrate heaters, process controllers and a sophisticated array of instruments for characterizing the vacuum.

The Shuttle Cross Bay Carrier remains in the Orbiter. It contains equipment to support the Free Flyer's independent operation, including an innovative communication system that routes commands and data in such a way that, from an operations standpoint, the Free Flyer appears to remain in the bay. This greatly simplifies communications and gives ground crews and Orbiter crews equal access to command and telemetry data.

On a typical mission, the Orbiter's remote manipulator arm grasps the unlatched Free Flyer, moves it outside the payload bay and releases it. Using its own propulsion,

the Free Flyer moves to a point 30-40 miles behind the Orbiter for about 50 hours of epitaxial thin film growing free of Shuttle contamination. The film growing cells are located on the wake side, or "clean" side of the disc; the "ram" side houses an avionics platform with 65 square feet of area for other experiments. On completion of the film growth experiment, the Free Flyer is recaptured by the Orbiter crew and, still in the grasp of the manipulator arm, it conducts additional research related to the near-Orbiter environment. Ultimately, the Free Flyer is returned to mate with the Carrier in the payload bay for return to Earth.

The WSF program contemplates four proof of concept missions. The first, WSF-1, was flown in February 1994. On that occasion, the Free Flyer remained attached to the manipulator arm and water vapor outgassing from the Orbiter contaminated the vacuum to a degree. Nonetheless, the mission was largely successful: WSF-1 generated a high wake vacuum and the epitaxy apparatus proved itself by growing five gallium arsenide single

crystal thin films, the first ever grown in space.

In July 1995, at *Spinoff* publication time, the WSF-2 was being readied for flight aboard the Space Shuttle. The mission plan called for growth of gallium arsenide and aluminum gallium arsenide films to demonstrate high electron mobility and advanced crystalline structures. The Free Flyer was to be deployed some 30 miles from the Orbiter, well out of contamination range, to form and characterize the first ultravacuum in LEO. **Above**, WSF-2 is pictured undergoing prelaunch checkout at Kennedy Space Center.

Two more flights are planned. WSF-3, targeted for 1996, is expected to show increased capability in the number and types of thin films grown, and in command and control of the growth process through a ground-based operations center. The program will advance another major step with WSF-4 in 1998; the WSF-4 Free Flyer will be equipped with solar panels for greater power and it will be capable of processing up to 300 wafers.

SMALL SATELLITE DEVELOPMENTS

NASA's Small Spacecraft Technology Initiative (SSTI) is a program designed to demonstrate technologies for reducing the cost and time of getting civil and commercial space missions from the drawing board to orbit. The program features new approaches to satellite design and development that will not only significantly lower cost, but will also permit the builder to incorporate commercial standards in the design and qualification process. Thus, the program will serve NASA's needs by allowing more frequent space missions within predictably lower budgets, and it will greatly enhance the competitive posture of U.S. space system manufacturers in the global market.

The program was launched in June 1994 with the award of contracts to TRW Space & Electronics Group, Redondo Beach, California and CTA Incorporated, Rockville, Maryland. Under the contracts, two industry teams will build small (600-900 pound) satellites ("smallsats" or "lightsats") to be developed within two years, then launched into orbit and serviced for an additional year. The small spacecraft are designed to accommodate a wide range of mission requirements through use of standard hardware and software adapted to various applications. The initial satellites were named "Lewis" and "Clark" after the leaders of the early 19th century U.S. expedition to the Pacific northwest.

Both companies have extensive experience in building smallsats. TRW, one of the nation's major aerospace manufacturers, developed smallsat technology for a number of military programs, built several lightsats for NASA, and is currently building commercial lightsats. CTA is among the world leaders in manufacturing small satellites systems, having built, launched and operated 21 lightsats, with nine others in development. Both companies head teams of experienced aerospace manufacturers.

Shown **at right,** the TRW Lewis satellite will be a 870-pound spacecraft carrying three payloads for Earth observation, commercial remote sensing and astrophysics. Where typically payloads have represented 10 to 20 percent of a spacecraft's total weight, TRW has established a goal of 50 to 70 percent.

Lewis will incorporate more than 25 new spacecraft and payload technologies for providing superior performance while minimizing cost and schedule time, among them miniaturized cryocoolers, advanced composite materials, faster data processors, lightweight propellant tanks and smaller star trackers.

TRW is building one of the three Lewis payloads, a hyperspectral imager that will observe Earth in 384 narrow spectral bands (the NASA Landsat has seven bands). Although Lewis is a technology demonstration spacecraft, it will also operate as a commercial remote sensing system, supplying Earth resources data to a broad range of industry participants and schools.

The other payloads aboard Lewis are a broad area sensor, developed by Goddard Space Flight Center, that will provide spectrally resolved Earth imagery, and an extreme ultraviolet spectrometer, built by the University of California at Berkeley, to take images of the night sky and the cosmic background.

The CTA/Martin Marietta Clark satellite **(lower right)** will demonstrate 36 advanced technologies, among them a mini-star tracker, a low-cost Sun sensor, advanced composites, shape-memory mechanisms, room temperature x-ray detectors, image

data compression, 3D imaging of atmospheric trace gases, and on-board data processing.

Clark will carry a modified commercial remote sensing imaging system and three science payloads. Principal focus will be on commercial remote sensing, which CTA sees as "the next frontier for domestic and international competition, with promise to become a multibillion dollar industry." Science research will include x-ray spectrometry, including solar flare measurements, cloud mapping and atmospheric tomography. CTA sums up its SSTI project as "a robust mission to deliver high value science, demonstrate multimission enabling technologies, use new design and qualification methods, and spawn commercially viable products in a two-year-to-flight profile."

SPACE RESEARCH LABORATORY

he middeck area of the Space Shuttle Orbiter's pressurized crew segment is the most cost-effective place aboard the Shuttle for conducting crew-tended scientific and commercial microgravity experiments. However, middeck locker space for experiments is very limited.

To meet a new for more experiment facilities per flight, SPACEHAB, Inc., Arlington, Virginia developed the SPACEHAB Space Research Laboratory pictured. Carried in the Orbiter's payload bay and accessed through an airlock, the SPACEHAB laboratory is a pressurized facility that roughly doubles the Orbiter's habitation volume and quadruples the volume available for crew-tended payload hardware. **At right,** the laboratory is shown in the Orbiter's payload bay on flight STS-63; the photo was taken by an astronaut perched high above the Orbiter on the remote manipulator arm.

Two of the basic SPACE-HAB modules were built and certified for Shuttle flight in a development program financed entirely with private capital. McDonnell Douglas Aerospace conducted design, development and construc-

tion of the modules under contract to SPACEHAB, Inc. SPACEHAB pays NASA for launch services and leases experiment space to U.S. and international private industry, universities, research institutions, and government agencies, including NASA.

The laboratory has flexible accommodations wherein payloads can be carried in racks and lockers and externally on the flat rooftop. Equipped with the necessary subsystems for power, environment control and life support, communications and data systems, the basic module is 13.5 feet in diameter, nine feet long, has 1,100 cubic feet of pressurized volume and a 4,800-pound payload capacity.

The SPACEHAB Space

Research Laboratory made a successful Space Shuttle debut aboard STS-57, Orbiter *Endeavour*, in June 1993. On that occasion, the module carried 22 experiments sponsored by more than 25 U.S. companies, NASA, a network of universities and research organizations, and the European Space Agency.

In February 1994, the laboratory went aloft on its second flight aboard STS-60, Orbiter *Discovery*. Once again it demonstrated successful performance, housing 13 commercial development experiments, including crystal growth research aimed at development of improved semiconductors, and an investigation of the structure of human insulin crystals directed toward possible

improvement in the treatment of diabetes.

SPACEHAB's third flight — also successful — was on STS-63, Orbiter *Discovery*, in February 1995 (see page 16). On SPACEHAB-3, SPACEHAB, Inc., introduced innovations to reduce significantly the demand on crew time: a video switch for trimming the time astronauts spend controlling video equipment and an experiment interface in the SPACEHAB telemetry system that reduces crew time in downlinking experiment data. Another new feature is a pair of 12-inch diameter windows on the roof of the module; one window has a NASA docking camera to assist in Shuttle/*Mir* rendezvous docking.

SPACEHAB, Inc. has developed a "double" module to increase its payload-to-orbit capacity and to provide greater flexibility in a mixed configuration, the forward half for locker and rack science experiments and the aft half for cargo/supply support of Shuttle/*Mir* missions.

Shown **above,** the double module is capable of carrying 10,000 pounds of cargo (experiments and supplies), but it will be constrained to 6,000-pounds on Shuttle/*Mir* flights because the dockings take place in *Mir's* high inclination orbit. The double module consists of one of the standard single modules joined together with the SPACEHAB Structural Test Article that was used to qualify the single module. The Italian company Alenia Spazio is preparing the double module and certifying it for flight. It will debut on Shuttle/*Mir* Mission 4 (STS-79, July 1996) and will also be used on *Mir* Missions 5 and 6.

RECOVERABLE SATELLITE SYSTEM

A new space system known as METEOR offers significant enhancement of the U.S. capability for commercial experimentation in space. METEOR is a compression of Multipurpose Experiment Transporter to low Earth Orbit and Return, a title that pretty much defines the project: demonstration of an unmanned commercial space system capable of flying and recovering small research and commercial payloads that require longer exposure to the microgravity environment than is possible aboard the Space Shuttle.

METEOR was developed and built by EER Systems Corporation, Vienna, Virginia, using components from a prior NASA program known as COMET. At *Spinoff* publication time, the METEOR system was being readied for its initial launch in midsummer 1995 from NASA's Wallops Flight Facility, Wallops Island, Virginia. METEOR's booster is the Conestoga launch vehicle, also developed by EER. Goddard Space

remainder by private industry — housed in two payload modules. Experiments requiring long duration exposure to space are housed in the service module, which is designed to transmit data from low Earth orbit for about a year. Experiments that are enhanced by periods of microgravity longer than the Shuttle can provide, and those that require postflight analysis on Earth, are housed in the recovery module, which was to remain in orbit for approximately one month, then return to a water landing about 100 miles off Wallops Island.

The experiment complement for the initial flight cut across a number of research disciplines, including navigational aids, ultraviolet and remote sensing instruments, materials exposure to space, plant/crystal growth, and new spacecraft technology demonstrations. The initial flight also marked the first launch of the Conestoga vehicle.

The METEOR system offers an affordable alternative for conducting in-space experiments that do not require human presence. Its launch by expendable boosters gives experimenters flexibility in selecting orbital parameters different from those of the Space Shuttle.

Flight Center is program manager.

The initial flight contained several microgravity experiments — about half sponsored by NASA and the

FIREFIGHTING TECHNOLOGY

Firefighting and fire prevention are areas of activity that seem to be special beneficiaries of aerospace spinoff technology. In recent years, for example, aerospace technology has been productively transferred to such civil use applications as portable firefighting modules; protective outergarments for workers in hazardous environments; a broad range of fire-retardant paints and foams; fireblocking ablative coatings for outdoor structures; shearing tools for rescue work (see page 62); and a number of types of flame resistant fabrics for use in the home, office or public transportation vehicles.

The City of Chicago is taking advantage of NASA's expertise in this area. Under an agreement concluded in April 1995, NASA and the Chicago Fire Department (CFD) will work together to apply aerospace technologies to improving CFD's firefighting and other emergency services.

A number of candidate activities have been identified for joint study. CFD's principal interest is in developing a personnel locator system that will enable on-scene fire officials to locate, track and, if necessary, rescue firefighters within a 2,400-foot radius of operations.

Another area of CFD interest is adapting NASA dynamic structural analysis techniques to determine if a structure is in imminent danger of collapse. Studies have indicated that there is a "signature" change prior to collapse of a building and it might prove feasible to develop a device that could monitor the signature in real time, providing fire officials a means of determining when firefighters should be evacuated from a troubled structure for their safety.

CFD is also interested in developing a new, portable, firefighter's air-breathing apparatus that would be less costly, lighter and allow longer use time. NASA and CFD will explore the possibility of applying space-derived liquid air technology to an advanced breathing system.

Other research possibilities include identifying the location of an emergency 911 call made from a cellular phone, and developing the capability to warn hearing-impaired drivers of an approaching emergency vehicle. MSFC is already working on such an Emergency Vehicle Alerting System (EVAS), a device that would assist drivers with hearing loss and those operating in a high-noise environment — a school bus full of children, for example. The EVAS would not only alert the driver, it would also provide information on the emergency vehicle's distance and direction of approach.

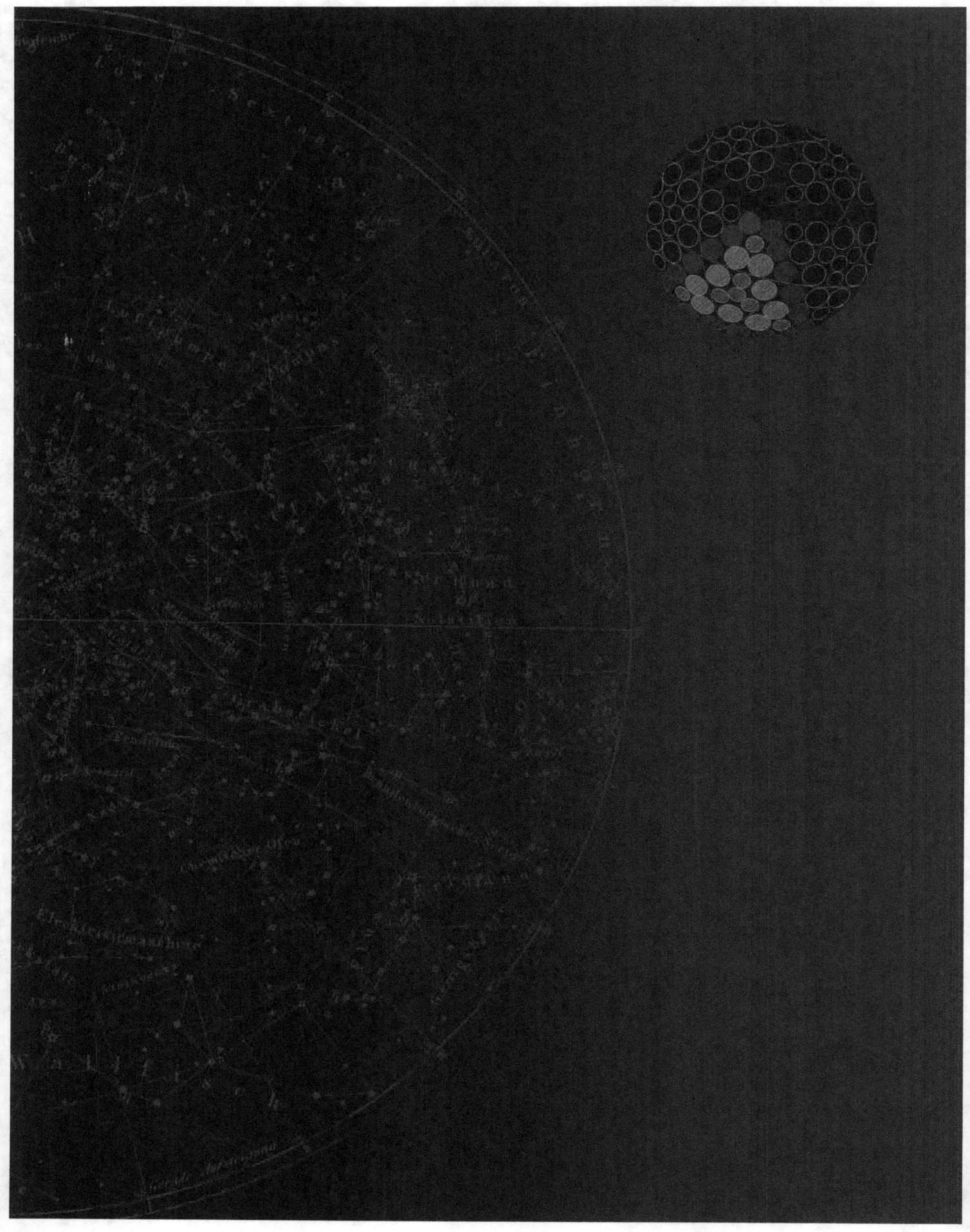

TECHNOLOGY TWICE USED

A representative selection of new products and processes adapted from technology originally developed for NASA mainline programs, underlining the broad diversity of spinoff applications and the social/economic benefits they provide

A dramatic advance in aviation hazard prediction heads a selection of technology transfers for improved public safety

On November 30, 1994, Continental Airlines Flight 1637, a Boeing 737-300 jetliner, took off from Washington (D.C.) National Airport bound for Cleveland. It was a routine, regularly scheduled flight — but to aviation safety officials all over the world it was something more; it was a historic moment that marked the introduction to commercial airline service of an instrument for detecting and predicting windshear, a leading cause of aircraft accidents.

The instrument was the Bendix RDR-4B airborne weather radar developed by AlliedSignal Commercial Avionic Systems, Fort Lauderdale, Florida. The first such system to be certified by the Federal Aviation Administration (FAA), the RDR-4B forward-looking radar is the product of a decade-long FAA/NASA/industry/academia research program spearheaded by Langley Research Center, which developed the technology base that enabled commercial manufacture of the AlliedSignal and other windshear detection/prediction systems.

Windshear is a sudden shift in wind direction and velocity. Its most violent characteristic is the microburst, a column of air which cools rapidly in a thunderstorm and generates intense downdrafts that create a maelstrom of powerful, swirling wind currents near ground level. Low and slow aircraft, such as airliners landing or taking off, are particularly vulnerable to the microburst, which can cause an airplane to lose aerodynamic lift and flight speed, and to plunge into the ground before the flight crew can take corrective action. That has happened on numerous occasions; windshear has been identified as the cause of more than 30 major aircraft accidents and the loss of hundreds of lives.

One of those accidents occurred at Dallas-Fort Worth Airport on August 2, 1985, when a powerful microburst caused the crash of a Delta Airlines Lockheed L-1011 jetliner with 137 fatalities. That accident triggered a Congressional mandate to the FAA to initiate a research/training effort toward curbing the windshear hazard.

In 1986, the FAA and NASA launched a joint program to develop the essential technology for detecting and avoiding microbursts. The FAA undertook an aircrew training program that focused on windshear recognition and procedures for recovering from its effects. FAA also led the development of ground-based windshear detection instruments, including the Terminal Doppler Weather Radar (TDWR) now being installed at 45 major U.S. airports; developed by Raytheon Corporation, the TDWR can accurately measure wind velocities in terminal areas and generate real-time aircraft hazard displays that are updated every minute.

NASA's Langley Research Center focused on development of airborne systems capable of predicting windshear. Another type of system, the reactive system, processes data from standard aircraft instruments to determine the presence of windshear. The reactive system only advises a pilot that he is in a windshear event, allowing him to increase engine power and possibly escape the hazard; however, the airplane might not be capable of recovering from a severe windshear at that point. Langley concentrated on a predictive system in the cockpit that would provide 10 to 40 seconds of warning, thereby enabling the pilot to determine the proper maneuver, add power for flight stability or avoid the windshear area altogether.

Augmented by technology application experts from Research Triangle Institute, Research Triangle Park, North Carolina, the Langley team initiated development of three different types of microburst sensors. The one that ultimately became the first in airline service is the Doppler microwave radar, which sends a radio wave ahead of the aircraft to "bounce" off rain-drops in the thunderstorm and return to the instrument; computerized measurement of the Doppler shift (the difference in wavelength frequency between the outbound wave and the returning signal) provides an indication of windshear velocity.

A second type of system — LIDAR, for Light Detection and Ranging — operates under the same Doppler shift principle but employs a laser beam instead of a radio wave. A third type, a passive infrared sensor, is based on the fact that a microburst is usually cooler than the sur-rounding air, thus can be detected by infrared measurement of the temperature differential ahead of the airplane. The infrared system is no longer in development; the LIDAR is still being investigated for possible use as a component of a future hybrid system or a multifunction weather/turbulence detection system.

Langley's task was essentially to build a technology base that would enable manufacturers to develop their own commercially viable, proprietary systems. The enormous job began with characterizing the windshear hazard and determining the warning time required. Extensive computational simulations, using computer models thoroughly validated by actual observa-tions, documented the structure, strength and evolution of microbursts from the smallest to the largest (about four kilometers). This work established the basic specifications for sensors and enabled development of algorithms for rejecting ground "clutter" that could confuse sensor signals. Langley also developed the all-important "F-Factor," an index that relates atmospheric motions to aircraft performance capability and assesses the degree of hazard to an airplane in a microburst encounter.

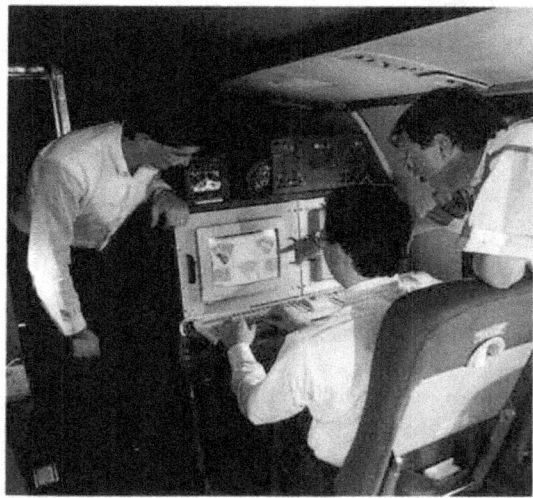

All of this knowledge gave manufacturers a broad knowledge base on how to extract windshear information from a sensor signal, how to process the data against hazard crite-ria, and how to alert flight crews to valid threats while rejecting "nuisance" indicators.

(Continued)

57

PUBLIC SAFTEY

In the 737's windowless research cockpit, aft of the regular flight deck, researchers study the special informational displays generated by ground and airborne windshear detectors.

By 1991, five years into Langley Research Center's windshear sensor development program, the technology had advanced to the point where validation of the sensors required actual flight tests in windshear conditions. For that job, Langley outfitted its unique flying laboratory with three types of sensors: the Doppler radar, the Doppler LIDAR and the infrared remote temperature sensor.

Formally known as the Transport Systems Research Vehicle (TSRV), the airplane was a preproduction model of the Boeing 737-100 jetliner equipped with a research cockpit in what would have been the forward section of the passenger cabin. Most of the sensor testing was conducted from this windowless "second cockpit," which was fitted with a full suite of flight controls and aircraft instrumentation, plus the sensor-related displays and data recording equipment. The TSRV's dual-cockpit configuration was designed to allow testing of new equipment and technologies in a safe environment (the pilots in the standard forward cockpit could take control of the aircraft should a malfunction occur in the research cockpit's instruments or equipment).

Langley selected two atmospherically-different field sites for the joint NASA/FAA flight test program: one at Orlando, Florida, the other at Denver, Colorado; both areas were noted for frequent microbursts in summertime. The flight test plan was simple: when ground radars detected windshear, the Lang-

Langley Research Center's Boeing 737 research aircraft is fitted with a Doppler radar windshear detection system (antenna shown) that sends a beam well ahead of the airplane to detect microbursts, evidenced by sudden large changes in raindrop velocities.

ley crew would scramble, take off, fly directly toward and into the microburst, observe and record the sensors' findings, then validate them by cross-checking with ground radar data and with data from an airborne reactive system for measuring windshear velocities in situ. Orlando flights were supported by a Terminal Doppler Weather Radar (TDWR) operated by Massachusetts Institute of Technology's Lincoln Laboratory; at Denver, a research radar of similar capability was operated by the National Center for Atmospheric Research. For safety purposes, the TSRV was flown into windshear at speeds higher than those of a normal jetliner approach and at altitudes greater than takeoff/landing levels, speeds of 240-260 miles per hour and altitudes of 750 to 1,500 feet.

Over the summers of 1991 and 1992, the Langley team conducted 130 flights and experienced 75 windshear events. The test program demonstrated that Doppler radar systems

This view from a tail-mounted camera shows Langley's 737, vectored by a ground radar, heading into a thunderstorm that may harbor microbursts. In a two-year program, the Langley team conducted 130 such flights and encountered 75 microbursts, testing the ability of NASA-developed sensors to accurately measure windshear velocities and provide 20-40 seconds of warning. **Below** is a closeup of a sensor alert showing potentially hazardous microbursts 2-3 miles ahead of the plane; the 737's position is at the bottom of the cone, dark red areas indicate greatest shear intensity.

offered the greatest promise for early introduction to airline service. The NASA forward-looking Doppler radar detected windshear consistently and at longer ranges than other systems, and it was able to provide 20-40 seconds warning of upcoming microbursts. The predictive sensors showed good correlation with data from ground radars and the on-board reactive systems.

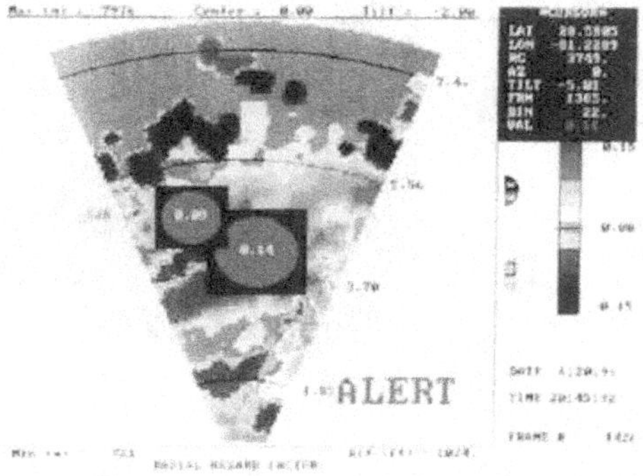

NASA's technology development effort was essentially complete but the Langley group continued working in a consulting capacity on the matter of FAA certification. No certification standards existed; they had to be invented and the Langley researchers now represented the world's most knowledgeable body of windshear expertise. Langley worked with the FAA and industry to develop a set of standards for certification of windshear sensors; collectively, the standards define the hazard, the cockpit interface and alerts to be given to flight crews, a suggested methodology for certification, and the requisite sensor performance levels. The NASA research was the basis for most of the specifications.

(Continued)

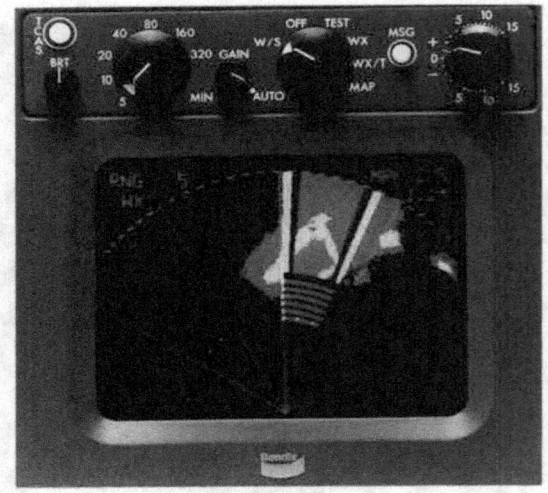

In 1990, the Federal Aviation Administration (FAA) ordered that all commercial carriers install windshear detection devices on their aircraft by yearend 1993. Reactive systems, which detect the presence of windshear without warnings, were being produced by several manufacturers and most airlines opted for that type of system. For those airlines who were already testing predictive systems and preferred to wait for their certification, the FAA offered an extended deadline of yearend 1995.

Meanwhile, NASA was trying to effect the broadest possible transfer to industry of the safety-enhancing predictive windshear technology. In addition to the usual practice of disseminating technical papers, NASA and FAA jointly sponsored well-attended windshear conferences and invited potential manufacturers to track the development of windshear detectors. Three major avionics manufacturers — AlliedSignal; Westinghouse Electronic Systems Group, Baltimore, Maryland; and Rockwell Collins Commercial Avionics, Cedar Rapids,

AlliedSignal's Bendix RDR-4B system includes the antenna (background), a receiver/transmitter (center) and a Planned Position Indicator (PPI). Shown in closeup **above,** *the PPI displays the direction and distance of a windshear microburst in addition to the normal en route weather display.*

Iowa — sent engineering teams to Langley to meet directly with NASA's radar engineering personnel and follow Langley's developmental effort step by step. The three companies each requested and were provided Langley's windshear simulations, which they used extensively in developing their own commercial systems.

On September 1, 1994, AlliedSignal's Bendix RDR-4B became the first predictive windshear system to gain FAA certification for airline operations. Three major U.S. airlines selected the RDR-4B, United and Northwest in addition to Continental; collectively they ordered more than 1,000 units. The technology is also being extended to foreign airlines; among those who have purchased the RDR-4B are Swissair, Alitalia, Iberia, Gulf Air and Kuwait Airways.

Westinghouse, a leading manufacturer of radar systems for 50 years, is producing two windshear detection/prediction versions of its MODAR (Modular Aviation Radar). For commercial air transport service, Westinghouse and Honeywell Inc. are jointly marketing the MR-3000 forward-looking Doppler radar. The companion MR-4000, known in the Air Force as AN/APN-241, is designed for military tanker/transports and was initially produced for installation in the Air Force C-130 transport by Lockheed Aeronautical Systems Company. In addition

to general weather information and windshear prediction, the MR-4000 offers advanced mapping and navigational capabilities. The windshear detection elements of both the MR-3000 and the MR-4000 were derived from NASA technology.

Rockwell Collins' Air Transport Division was a participant in the Langley test program; under NASA contract, the company supplied the hardware for the Doppler radar tested, first flown in July 1990. Collins later developed its own algorithms and incorporated them into a modified version of the company's WXR-700 weather radar, known as the WXR-700WR forward-looking windshear radar. The predictive windshear system, tested aboard a Continental Airlines 737 and Collins' own Gulfstream I, is designed for installation in new aircraft or retrofit in aircraft already in service. Collins initiated system deliveries to airline customers in 1995.

It took almost a decade to bring the predictive windshear system from concept to commercial availability, but that, aviation experts say, was a remarkably brief period, considering the complexity of the phenomenon and the virtually non-existent knowledge base at the outset of the program. The program stands as a model of cooperative endeavor by a broad segment of the U.S. aviation community, including government agencies, aircraft manufacturers, sensor manufacturers, airlines, research organizations and academia.

Engineers of Collins Commercial Avionics work on windshear displays in the company's laboratory mock cockpit facility. The product, Collins' WXR-700WR predictive windshear system, is shown at right; at bottom is the avionics box that houses the computer and above it is the radar disc, which is mounted on a movable arm in the nose of the airplane.

RESCUE EQUIPMENT

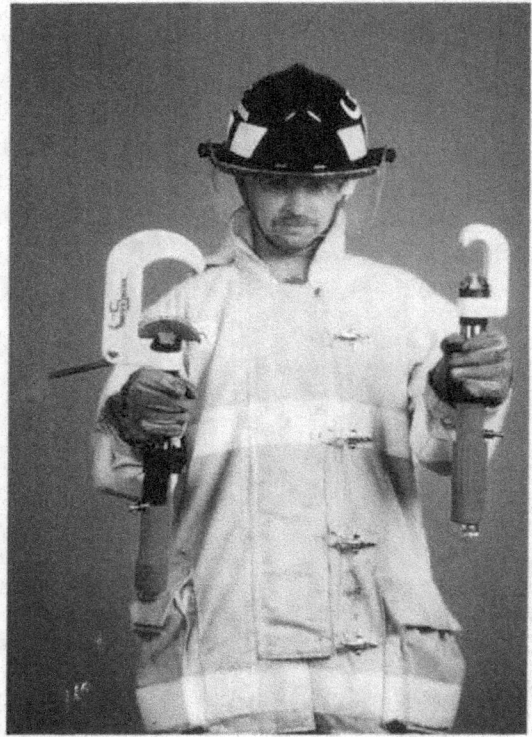

Below, a firefighter of the Torrance (California) Fire Department is displaying a new generation of lightweight portable emergency rescue cutters for freeing accident victims from wreckage. Known as Life-shear cutters, they incorporate NASA pyrotechnical separation technology and were developed under a cooperative agreement that teamed NASA and Hi-Shear Technology Corporation of Torrance.

The cutters were developed under the Clinton Administration's Technology Reinvestment Program (TRP), an effort to transfer government-sponsored technology to the U.S. commercial marketplace. The development project was undertaken to meet the need of some 40,000 U.S. fire departments for modern, low-cost emergency cutting equipment.

Prior cutting equipment employed expensive gasoline powered hydraulic pumps, hoses and cutters for use in accident extrication. To eliminate much of this cumbersome equipment, the Lifeshear design team opted for pyrotechnically-actuated cutters, which made possible a 50 percent weight saving and a 70 percent reduction in cost. The key technology is a modified power cartridge, a miniature version of the cartridges that actuate pyrotechnical separation devices aboard the Space Shuttle and other NASA spacecraft. Hi-Shear Technology Corporation is an industry leader in the development and manufacture of pyrotechnically actuated thrusters, explosive bolts, separation nuts, pin pullers and cutters and has supplied such equipment for a number of NASA deep space missions plus the Apollo/Saturn, Skylab and Space Shuttle programs. Although the company had broad experience in pyrotechnics, it lacked hands-on experience and training in vehicle rescue, therefore asked the City of Torrance Fire Department and Fire Chief Scott Adams to contribute their user experience to the development effort.

Hi-Shear offers the Lifeshear cutter in the two versions shown in **at right above,** the LS 100 (top unit) and the LS-200 (lower unit). The LS 100 has a 3½-inch "jaw" opening, is 26 inches long and weighs 11½ pounds. The LS 200 has a 1½-inch opening, is 22 inches long and weighs only 8½ pounds. The tools take only about 30 seconds to set up and they require no pumps or hoses; they can sever automotive clutch and brake pedals or cut quickly through roof posts and pillars to remove the roof of an automobile. They provide a bonus to firefighters and rescue personnel in that they reduce the probability of lower back injuries that occur with the use of hydraulically-powered equipment.

The accompanying photos illustrate two typical applications of the cutter. **At right,** a

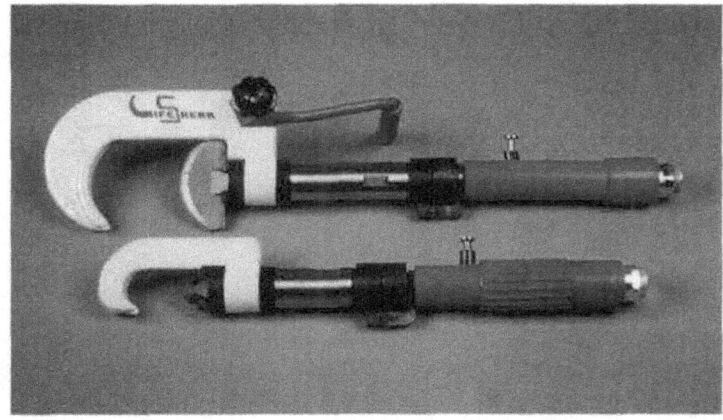

firefighter is clearing an egress route by cutting through reinforcement cable and bars in a collapsed-structure situation. **Above right,** a firefighter is cutting off the roof of a car to provide access to a trapped victim.

Priced at less than $2,000 per unit, the new equipment is cost-effective for the smallest fire departments and rescue squads and its easy transportability to remote areas makes it attractive for military/civil helicopter

search and rescue operations. Since the Lifeshear cutters were introduced in 1994, they have generated wide interest among other possible users, such as law enforcement agencies, private utility companies, mining interests, and elements of the Department of Defense.

Managed for NASA by Jet Propulsion Laboratory, the Lifeshear development effort was accomplished with funding of $1.6 million, shared by NASA and Hi-Shear. The contract award, made in February 1994, was the first of the national TRP cooperative agreements. Hi-Shear was able to bring the cutters to production and commercial availability in only six months, and by mid-1995 the company was manufacturing 600 units a month. The Lifeshear program supports seven machine shops in the Los Angeles area and employs 40 people.

TRAFFIC MONITOR

Below, Craig Sullender of Intelligent Vision Systems, Inc. (InVision™), Houston, Texas is running a test on the company's TDS-200 Traffic Detection System, which is being produced commercially and is in operational service at various locations in the U.S. The TDS-200 incorporates NASA technology originally developed for Earth observing imaging satellites.

Traffic monitoring is a difficult task requiring a wide range of sensors and functions that must operate under rigorous environmental conditions. Monitoring devices range from simple push buttons to very complex computer based systems. InVision's approach involved development of a multipurpose unit capable of handling the whole range of sensing, from the simplest task to very complex functions.

Sensing is performed by identifying shapes (vehicles, pedestrians), detecting movement or the lack of it, counting individual objects in their respective lanes and calculating their speed; information of this type is important to highway control engineers. The sensor employed in the system **(top right)** uses NASA CCD

(charge coupled device) technology in which silicon chips convert light directly into electronic or digital images, which can be manipulated and enhanced by computers.

The InVision system's sensors are mounted above the traffic on poles or span wires, enabling two sensors to view a standard intersection. A special feature of the TDS-200 electronically compensates for "swing and sway" movement and rotational error of the sensors; the sensor can swing and still identify and differentiate among moving and fixed objects in their correct positions.

A proprietary combination of electronic shutter and gain control provides exceptional dynamic range from extremely low light to direct sunlight with excellent picture clarity. Sensor output provides sig-

nals to the image digital signal processor (DSP), still frame video and optionally live video.

The TDS-200 selects several light wavelengths of the viewing area, processes the images and stores the information. The DSP performs several processing tasks concurrently in real time. Problems related to shadows, adjacent vehicles and wet street reflections are eliminated. The fixed background of the image is removed so that only changed or detected objects remain for processing. The photos **at right** display objects as "seen" by the computer after processing. A company-developed tracking algorithm, coupled with high speed parallel processing, enables tracking all objects in the field of view.

InVision president Paul Mayeaux credits NASA technology with the breakthrough

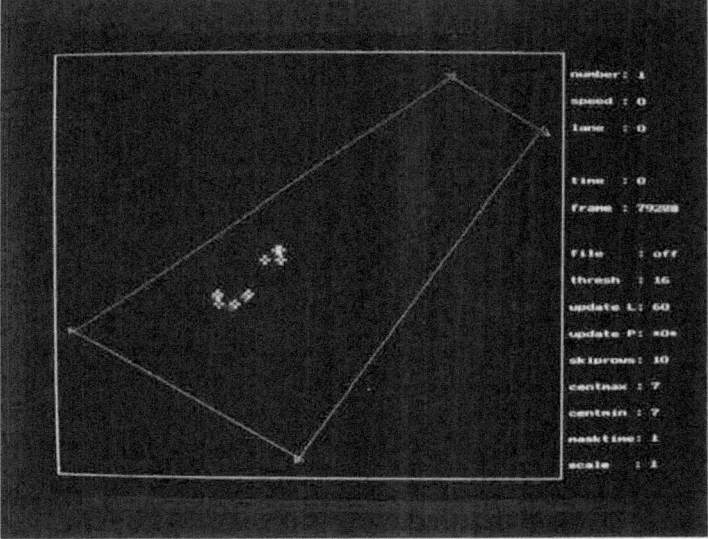

that solved the principal problem in the development: getting reliable image recognition in rain, fog and other bad weather. Says Mayeaux: "After three years of limited R&D success and dwindling enthusiasm, our research group (five partners sharing time and capital) realized that we had to find another image sensing approach. We had an economical computer, super signal processing hardware and software, but poor imagery. The problem was unreliable image acquisition in harsh environmental conditions."

In an issue of *NASA Tech* *Briefs* (see page 137), the researchers found an article describing a NASA technology developed for satellite imaging that utilizes multiple electromagnetic frequencies to improve image acquisition in all weather conditions. The InVision group requested and received from NASA a Technical Support Package containing more detailed information. The technology, plus advice and consultation provided by Johnson Space Center, proved to be the solution to the problem after modification of the technology to meet InVision's special design requirements. InVision subsequently applied technology from two other *Tech Briefs* articles that described related work accomplished by Jet Propulsion Laboratory and Lewis Research Center.

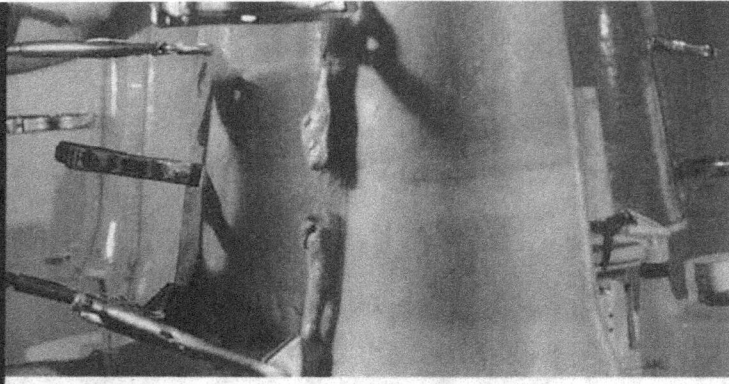

Heading a selection of spinoffs in health and medicine is a vision enhancement system derived from space technology

Some three million Americans suffer chronic visual impairments known to ophthalmologists as "low vision," which means that they are not totally blind but retain some degree of useful vision. These disabling impairments cannot be corrected medically, surgically or with conventional eyeglasses. They can, however, be corrected by a newly available video headset that offers people with low vision a view of their surroundings equivalent to the image on a five-foot television screen four feet distant from the viewer.

The headset is formally known as the Low Vision Enhancement System (LVES, pronounced Elvis). It will not make the blind see but for the majority of those in the low vision category it will ease such everyday activities as reading, watching TV, cleaning house, shopping, or working at jobs or hobbies.

LVES was introduced to the commercial marketplace in 1994 after almost a decade of development by a multiorganizational team that included NASA, the Wilmer Eye Institute of The Johns Hopkins Medical Institutions, the Department of Veteran Affairs, and Visionics Corporation, Minneapolis, Minnesota, which manufactures the system under license. The device was invented and patented by a trio of researchers: Dr. Robert W. Massof, director of Wilmer Eye Institute's Lions Vision Center; Dr. Thomas Raasch, also of Wilmer; and Dr. Donald O'Shea of Georgia Tech.

The Wilmer researchers, in cooperation with scientists of NASA's Stennis Space Center, used NASA technology for computer processing of satellite images, along with technology for head-mounted vision enhancement systems originally intended for space station use. The transfer of these technologies made it possible to improve the capabilities of low vision patients by appropriately enhancing and altering images to compensate for the patient's impairment.

The LVES consists of a head-mounted video display worn like goggles, a set of three video cameras, and a control unit that allows the user to select and control the cameras, and to adjust contrast and image polarity to suit the user's needs. The cameras feed the images to a computer that corrects for the particular vision problem of the user, then sends the images to the video display in the goggles.

Visionics sales literature succinctly describes the technology: "The LVES does for video images what headphones do for sound. It presents the enlarged image of a video screen to each eye in a way that creates a sense of being immersed in a video scene." The enlargement is accomplished by a patented system of optical mirrors, designed and manufactured by Polaroid Corporation, Cambridge, Massachusetts, that project the video images onto the wearer's retinas.

LVES is powered by a battery carried in a nylon belt pack along with a control unit that allows the wearer to adjust contrast and magnification. The belt pack, including the battery, weighs 2½ pounds and is about the size of a book. The headset, worn like aviator's goggles, weighs 34 ounces.

The system is used by people with a variety of low vision conditions, particularly those who have experienced loss of central vision, the part of vision normally used for reading.

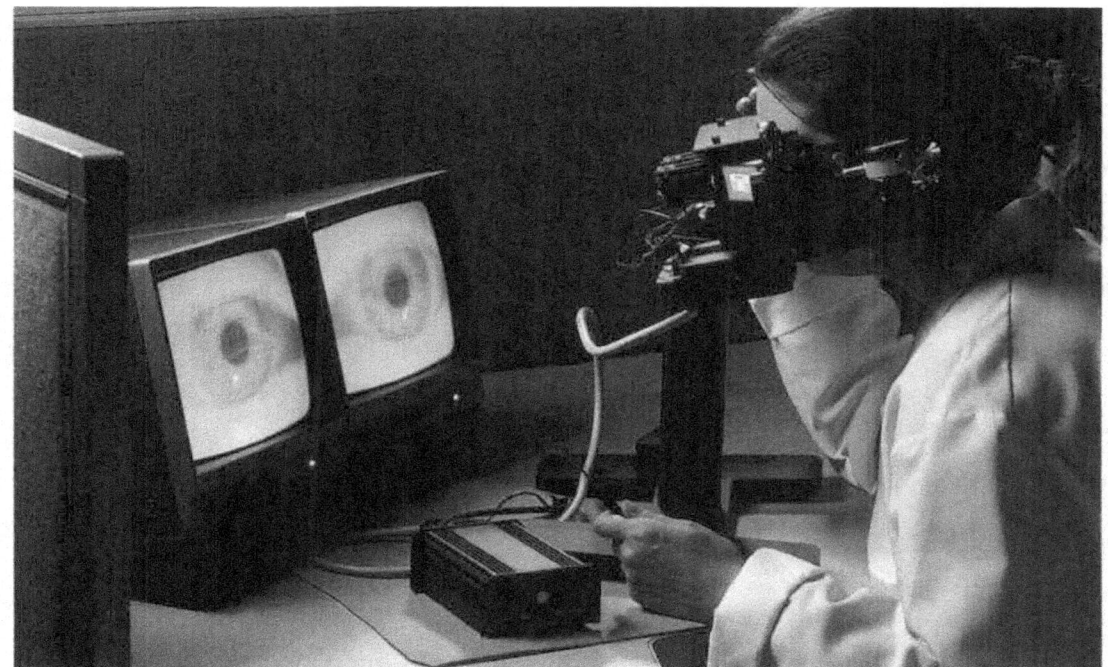

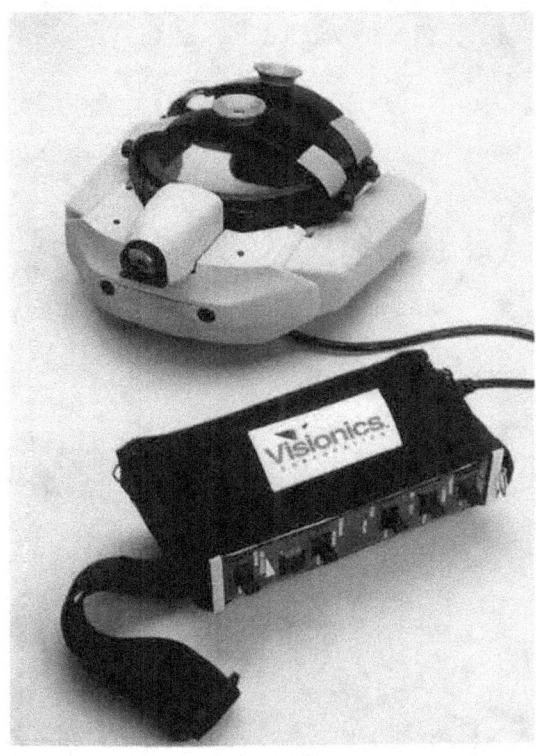

Shown above *is the Low Vision Enhancement System (LVES) produced commercially by Visionics Corporation. The upper unit is a head-mounted video system worn like goggles; the lower item is the control unit. The system corrects for vision defects and offers visually impaired people a clear view of their surroundings.*

These patients may have macular degeneration associated with aging or diabetic retinopathy, in which diabetes causes swelling and leakage of fluid in the center of the retina, the macula. The system also has benefited people who have lost peripheral or side field vision, a problem associated with glaucoma, an increase of fluid pressure inside the eye that damages the retina and optic nerve, and people suffering from retinitis pigmentosa, a progressive degeneration of the retina that results in tunnel vision and extreme sensitivity to light. Persons with optic nerve disease and congenital damage to the retina also use the LVES successfully.

The LVES development began in 1985, when Wilmer Eye Institute's Dr. Massof met with NASA officials to see if there was an emerging aerospace technology that might be adapted to low vision enhancement. Stennis Space Center and Wilmer then began development of a laboratory-based prototype system of real-time image processing. In 1992, the prototype LVES was unveiled and publicly demonstrated. Eleven centers of the Department of Veteran Affairs provided the patient population and conducted clinical trials of the system, on the basis of which final design modifications were incorporated into the headset. Visionics Corporation was founded in July 1992 to manufacture and market LVES and future improvements to the technology developed by Johns Hopkins.

STRESS PREDICTION SYSTEM

In the early 1970s, NASA contracted with the University of Michigan's Center for Ergonomics for development of a computerized biomechanical model capable of predicting stresses on various parts of the human body. Planning for possible advanced lunar missions and space station extravehicular activity, NASA wanted to know how astronauts' bodies would react under various gravitational pulls and space suit weights.

The Center for Ergonomics successfully developed an algorithm, or mathematical formula, for determining what type of stress and what degree of load a body could stand, taking into consideration the body's height, weight and posture. That technology has been commercialized with the ISTU™ (Isometric Strength Testing Unit) Functional Capacity Evaluation system **(right)**, manufactured by Ergometrics, Inc., Ann Arbor, Michigan.

The hand grips the subject is pressing measure the lift/push/pull force being applied. The horizontal bar slides upward and downward from overhead level to foot level; varying the bar level and the body position enables testing the lift capacity of a subject. Tasks such as lifting a heavy box from floor level, or lifting a weight overhead, or pulling/pushing a cart can be simulated. A computer printout provides the subject's test score and the exertion expended, and identifies the muscle group (knee, hip, ankle, etc.) that limits the subject's performance on a particular task.

Biomechanical strength modeling compares the physical stresses generated in the

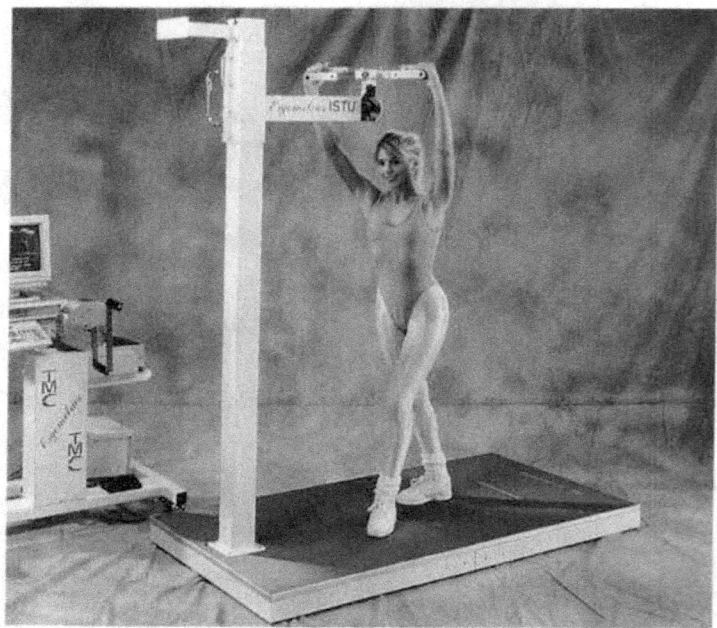

body with the resultant force capabilities of industrial workers; it translates job stresses into terms of human ability to handle them, and is therefore an effective tool of personnel evaluation, personnel selection and job redesign.

The isometric strength test is useful in disability evaluation; for example, it shows how much an injured worker is capable of lifting, an aid to disability rating, return-to-work targeting or detection of malingering. It is also useful for preemployment examinations, identifying candidates capable of meeting predetermined physical requirements; this helps reduce the number of future injuries and therefore reduces workers' compensation claims. The system also has utility in redesigning unsafe jobs and in rehabilitation programs; strength measurement at the start of a therapy program and at intervals thereafter allows the therapist to determine if there has been actual improvement.

™ISTU is a trademark of Ergometrics, Inc.

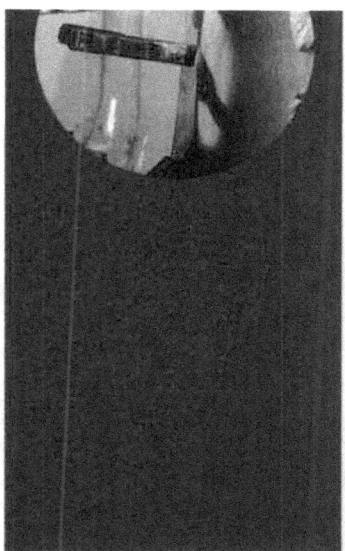

WEAR MEASUREMENT SYSTEM

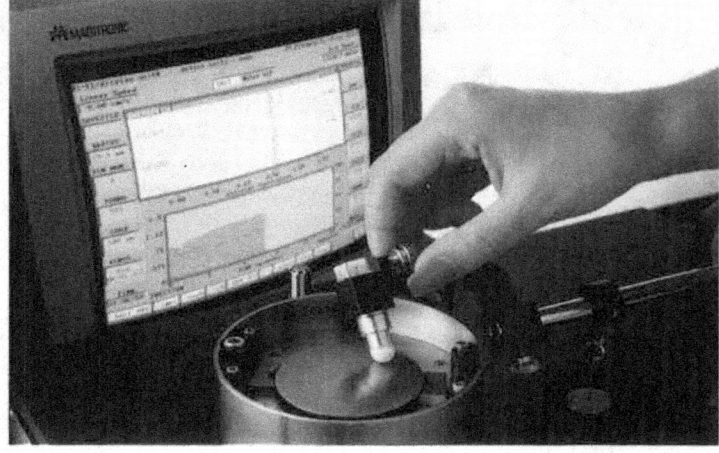

The ISC-200 Tribometer System is a device for measuring the wear on a material's surface. The system is marketed by Implant Sciences Corporation (ISC), Wakefield, Massachusetts, a company that provides a wide range of surface treatment and modification products and services, such as direct ion implantation, ion beam sputter coating and cathodic arc coatings.

The ISC-200 is a pin-on-disk type of tribometer. It somewhat resembles and operates much like a conventional record player. The pin-on disk test involves rotating the disk or pin specimen, which are in contact with each other (above); the rotation creates

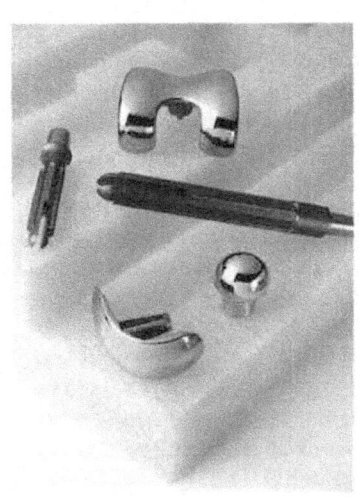

a wear groove on the surface of the disk. The system can be used to measure the coefficient of friction, the wear behavior between materials, and the integrity of thin films or coatings.

The ISC-200 Tribometer has biomedical applications, for example, measuring friction or wear on orthopedic devices, such as the parts, shown **at lower left,** used in knee and hip joint replacement, it also has utility in electronics manufacture, such as testing disk drives or tape drives, and in the automotive industry as a means of testing coatings on body parts or measuring wear on engine parts.

The ISC tribometer was originally developed by Lewis Research Center as an in-house wear test apparatus. ISC, working on a NASA

contract to develop coatings designed to enhance the wear capabilities of materials, had no way of testing the coatings they were developing. The company initially adapted the Lewis device to its own use, then converted the laboratory apparatus to a user-friendly commercial system that eventually became a complete line of tribometers. The first commercial versions were marketed in 1991; since then ISC has advanced the technology considerably; current systems can maintain a variety of simulated conditions, including controlled humidity, body temperature, high temperature and high vacuum.

PROSTHESIS MATERIAL

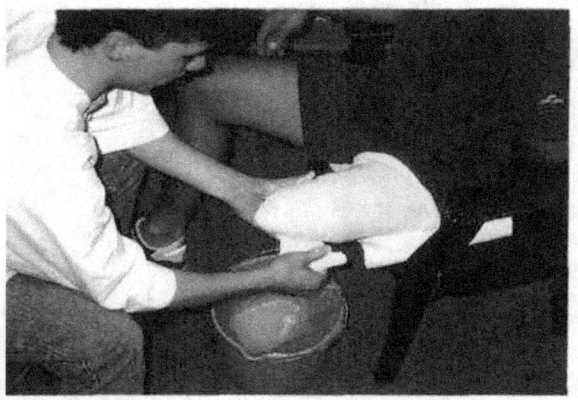

The Harshberger Prosthetic & Orthotic Center, Inc., Birmingham, Alabama offers personalized care to the injured, including crafting and fitting orthopedic appliances such as artificial limbs. The Center has a subsidiary company, also located in Birmingham, called FAB/CAM, which fabricates prosthetic components using computer-aided design and manufacturing (CAD/CAM) techniques.

Certified Prosthetist Jerald J. Harshberger, who heads the Center and the FAB/CAM subsidiary, learned of Marshall Space Flight Center's (MSFC) problem solving assistance to industry and paid a visit to MSFC. He outlined the problem: there was need to replace the plaster and corn starch materials FAB/CAM was using to produce master molds for prosthetic devices; the plaster molds were heavy, fragile, difficult to ship and store, and unrepairable if damaged.

MSFC's Technology Utilization Office provided a tentative solution. MSFC, in cooperation with Martin Marietta Manned Space Systems, New Orleans, Louisiana, manufacturer of the Space Shuttle External Tank (ET), was in process of creating a commercial derivative of the foam insulation system used to protect the ET from excessive heat. Some of the properties of the foam that made it effective as an ET insulator offered attractive commercial potential; the material was lighter, less expensive, stronger at high temperatures, more readily machinable and more stable than earlier foam insulators.

Martin Marietta was asked to produce a trial quantity of foam blanks for testing in a carving machine at FAB/CAM. The carver uses the blank for cutting a mold of a patient's residual limb; the mold is then used to form the socket for an artificial limb. The new material proved to be the solution to the problem.

Says Jerald Harshberger: "These foam blanks provided us with an alternative substitute in replacing plaster blanks. NASA foam is lighter, less expensive and faster to manufacture. Through NASA assistance, we were able to lower the cost and time of manufacturing, which financially helps our company and

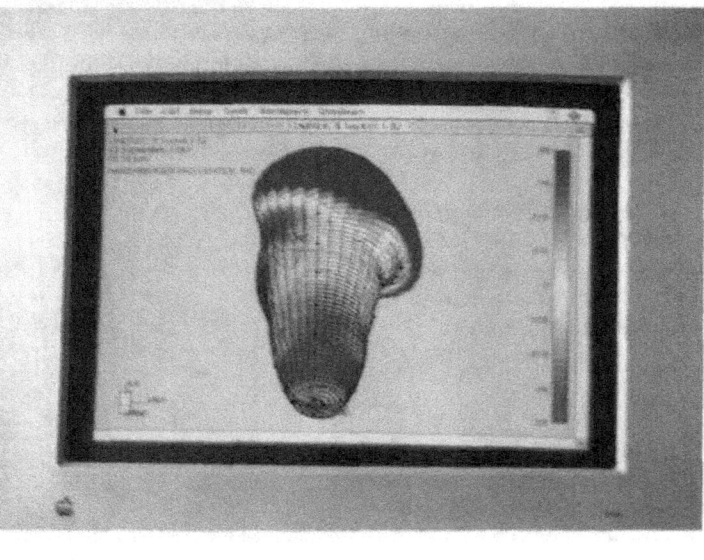

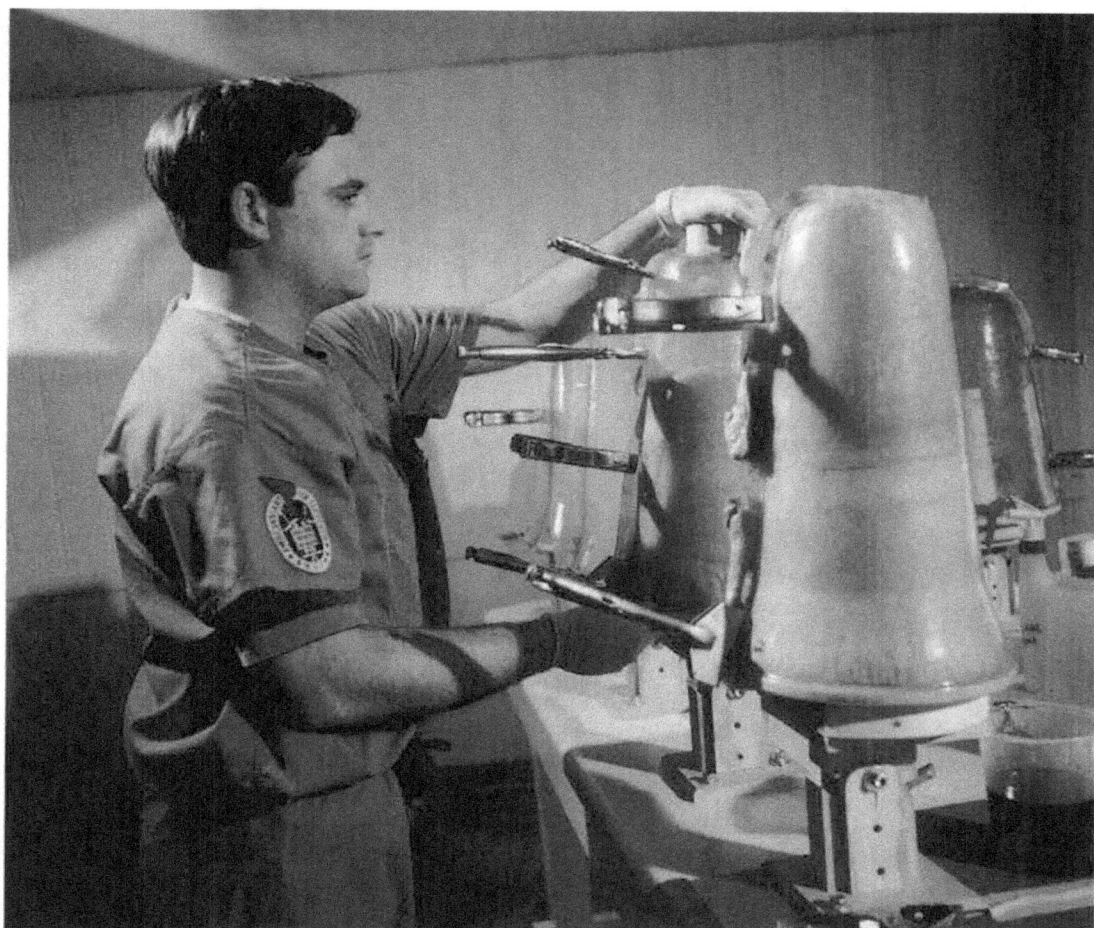

our patients. We are also able to mass produce the foam blanks and distribute them to other prosthetists across the nation."

At far left, a FAB/CAM prosthetist is taking an impression of a patient's residual limb. The digitizer of the CAD/CAM system "reads" the inner surface of the impression **(above opposite)** and the data is entered into the computer. The computer constructs a three-dimensional image of the appropriate socket **(bottom opposite)** and the prosthetist then modifies, designs and sculpts the socket shape on the computer's monitor. The foam blank is inserted in the computer-directed carver **(left below)** which carves a positive model. A socket is then "thermoformed" over the model and the patient is ready to be fitted.

This spinoff from Space Shuttle technology represents a dual technology transfer, because in addition to the benefit to the medical community it also provided Martin Marietta Manned Space Systems with a product that has commercial potential in high temperature insulation and structural applications as well as in prosthetic appliance manufacture. Martin Marietta is producing and marketing the foam system, initially for the prosthesis market, under the trade name MARCORE™. In

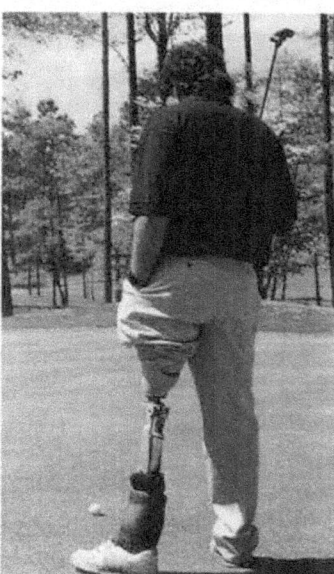

the **top photo,** a Martin Marietta technician is preparing an above-the-knee foam blank; **Above,** an amputee golfer models an artificial limb made with the MARCORE material.

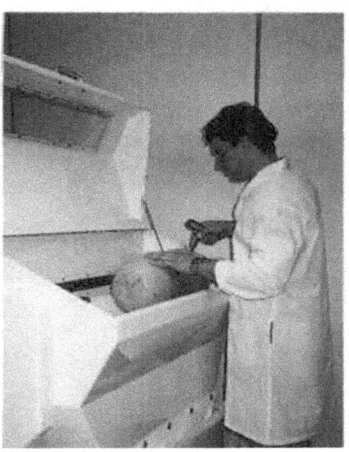

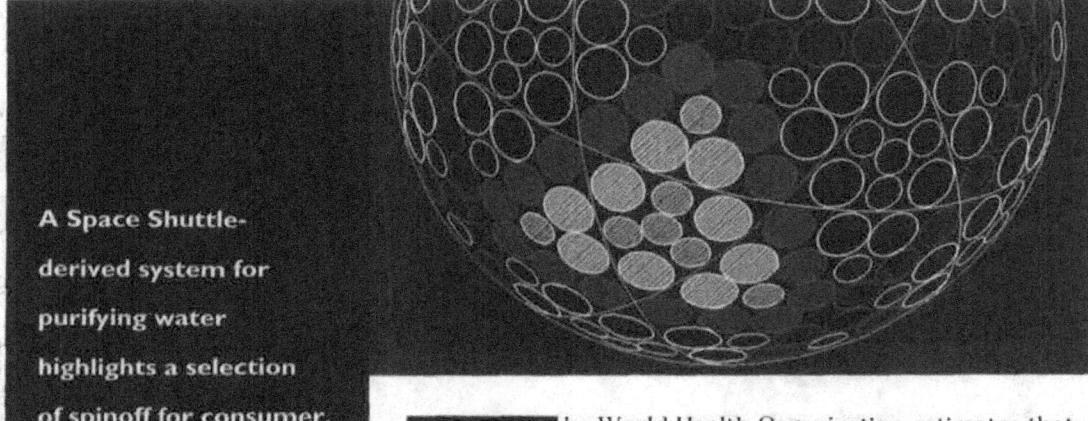

A Space Shuttle-derived system for purifying water highlights a selection of spinoff for consumer, home and recreational use

The World Health Organization estimates that more than one billion of Earth's people do not have daily access to clean, safe drinking water and millions of deaths each year are directly related to contaminated water. That underlines a long-standing humanitarian need for an effective yet inexpensive system of water purification that can be scaled up to municipal water treatment capacity for developing nations and can additionally serve as a backup emergency system in developed nations.

Such a system, a spinoff from Space Shuttle technology, is now commercially available and is, in fact, being installed in growing numbers in developing nations. Called the Regenerable Biocide Delivery Unit, it is a prime example of a dual-use technology, one developed for space application but also adaptable to a pressing Earth-use requirement. Its worldwide importance led to its selection as the 1993 NASA Government Invention of the Year and NASA Commercial Invention of the Year. The recipients of the awards were Richard L. Sauer, Water Quality Manager at Johnson Space Center; Gerald V. Columbo, vice president of Umpqua Research Company, Myrtle Creek, Oregon; and Clifford D. Jolly, also of Umpqua.

The Regenerable Biocide Delivery Unit uses iodine, rather than chlorine, to kill bacteria. Though new to the commercial market, the system represents an advanced phase of a technology that has been around for a considerable time. The effectiveness of iodine as a biocide has long been known, but its commercial application has lagged for want of an efficient long-term delivery system. The new invention solves that problem and thus opens up possibilities for a new line of iodination systems for portable, emergency, industrial and community use.

In operation, the unit disinfects water by passing it through a bed of iodinated resin known as MCV® resin; it gets its name from the cartridge that contains the iodine resin, the Microbial Check Valve. The MCV cartridge is not new; it was developed by Umpqua Research in the mid-1970s to purify the Space Shuttle Orbiter's supply of drinking water, and it has been used on every Shuttle flight since.

What is new is the ability to regenerate the iodinated resin bed. The original MCV cartridge was a limited life device that had to be replaced periodically with a new cartridge. The Regenerable Biocide Delivery Unit eliminates the frequent replacement requirement; it regenerates cartridges in place without removing them from the unit. Tests have shown that the cartridge can be regenerated more than 100 times.

It is this regenerable capability that makes the invention a significant advancement in water purification. Use of a regenerable resin bed answers the need for a practicable iodine delivery system. Use of iodine provides a most effective disinfectant and eliminates the hazards associated with chlorine. Regenerability also enables a sharp reduction in cost — ultimately perhaps less than one cent a gallon, compared with $1 - 2 per gallon for bottled water. That,

Gerald V. Columbo of Umpqua Research Company displays a Shuttle-derived Umpqua filter. The basic filter is filled with iodine resin, dispensed in a water supply for killing bacteria. The filter employs other purifying materials; the colored banding identifies the different materials used to remove different contaminants.

coupled with the Shuttle-derived characteristics of high reliability and low maintenance, and the inherently long shelf life of the system, make it attractive as an emergency backup system for use in developed nations during power outages, floods and other natural disasters, and as a water treatment system for whole communities in developing nations. The first large-scale application of the technology is already under way in Vietnam.

(Continued)

*MCV is a registered trademark of Umpqua Research Company.

The development of the Regenerable Biocide Delivery Unit evolved over a number of years, beginning in the mid-1970s when NASA was looking for an advanced method of purifying drinking water for the Space Shuttle Orbiter. A NASA contract to Umpqua Research, a small business specializing in air and water support equipment, resulted in development of the MCV (Microbial Check Valve) iodine-dispensing system. Introduced to Shuttle service in 1979 and still in use, the system proved reliable and effective for the relatively short-duration missions flown by the Shuttle Orbiter.

In 1989, looking to the International Space Station and future long-duration spacecraft, NASA awarded Umpqua a new contract for development of a system that would overcome the limitations of the original MCV system and allow continuous, controlled release of iodine over a lengthy period. That contract produced the Regenerable Biocide Delivery Unit, demonstrated in 1993 and being considered as the baseline water purification system for the International Space Station.

Umpqua Research was granted an exclusive worldwide license for commercial marketing of the technology. Umpqua, oriented to R&D and analysis work rather than manufacturing, is offering sublicensing opportunities to manufacturers of water filtration equipment. The principal sublicensee — the Vector Group of Companies — has embarked on a large-scale program to make clean water more widely available in developing nations.

A number of sublicensees are using the NASA/Umpqua iodinated resin technology for a variety of home-use and commercial products. In the photo is a sampling of home-use systems, a countertop model (left), an under-the-counter unit (center), and a faucet attachment unit (right). In the foreground is the original Shuttle Orbiter cylinder.

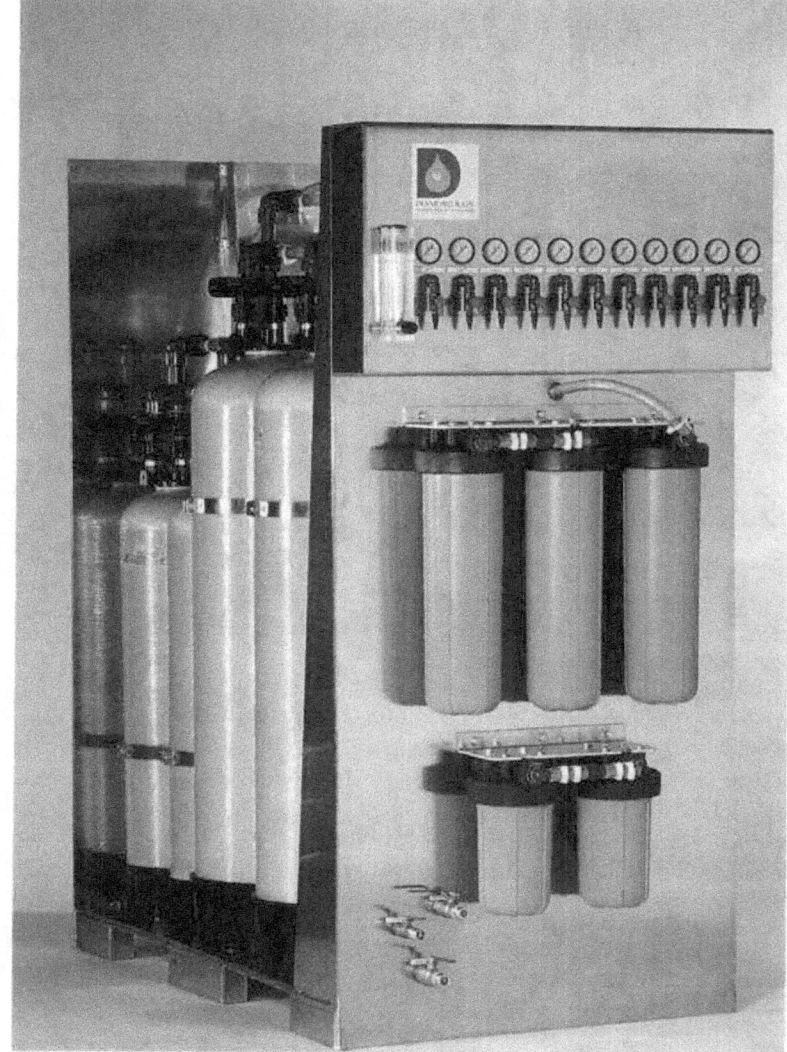

This eight-foot-long model is a representative DIAMOND RAIN water purification system, produced under Umpqua license by Vector Manufacturing Corporation in sizes ranging up to a unit that can cleanse drinking water for 35,000 people.

Represented in the U.S. by Vector Environmental Technologies, Sparks, Nevada, Vector is an international organization that focuses on technology for correction of such global problems as water contamination, marine oil spills, debris removal from waterways, and hazardous material storage.

Vector's manufacturing arm produces the DIAMOND RAIN Water Purification System, a modular system sized for any requirement from a single household to small community. The DIAMOND RAIN unit first passes raw water through a filtration stage to capture suspended particles, then through a granular activated carbon filter to absorb organic molecules — such as herbicides and petroleum contaminants — and to help remove unpleasant odors and taste.

The heart of the DIAMOND RAIN system is the Microbial Destruction Unit, an adaptation of the NASA/Umpqua technology that destroys pathogenic microbial organisms. In the final stage, the water passes through a series of post-filters, then is routed to a storage tank.

In July 1993, Vector signed an agreement with the Government of Vietnam for a water purification project of enormous scope: installation of 10,000 DIAMOND RAIN units over a five to six year period. The program focuses initially on making systems available to rural communities, hospitals, clinics, schools, hotels and households. The first units were shipped in 1994. Eventually, facilities for manufacturing DIAMOND RAIN systems will be established in Vietnam under an agreement among Vector, the Vietnamese government, and a state-owned company that oversees the country's water infrastructure.

Vector is also engaged in a joint venture with a Saudi Arabian firm for marketing DIAMOND RAIN systems in Saudi Arabia, Kuwait, Oman, Bahrain, Qatar and the United Arab Emirates. Additionally, Vector is introducing the water purification technology in the Philippines, Sri Lanka, Mexico, Egypt and Angola. Ongoing discussions with a number of countries suggest that DIAMOND RAIN systems will find extremely broad applications in Asia, Africa and Latin America.

GOLF AERODYNAMICS

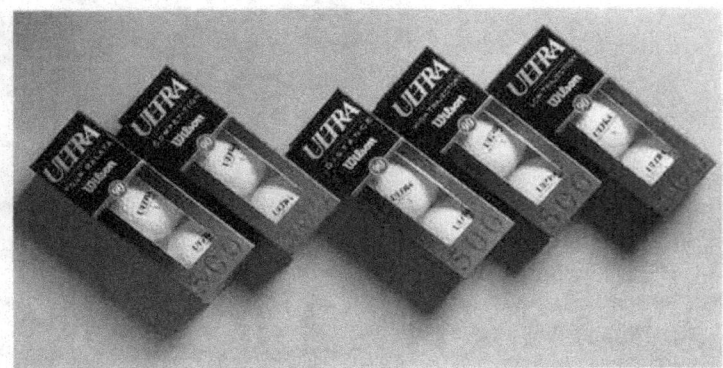

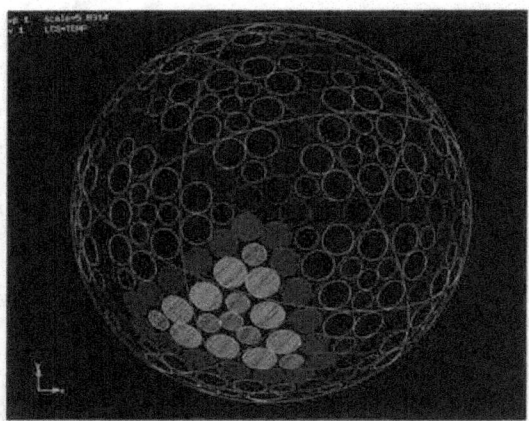

The Ultra® 500 Series golf ball **(right)**, introduced in 1995 by Wilson Sporting Goods Company, Humboldt, Tennessee, has 500 dimples arranged in a pattern of 60 spherical triangles. That probably has little meaning unless you are a golf pro or a low handicap amateur, but the design represents a major departure from Wilson's traditional ball-design approach. It is a design that employs NASA aerodynamics technology to create what Wilson Golf calls "the most symmetrical ball surface available, sustaining initial velocity longer and producing the most stable ball flight for unmatched accuracy and distance."

Seeking to optimize both distance and accuracy, Wilson Golf assigned a design engineer formerly associated with NASA programs to conduct extensive research and testing on dimple patterns. The result of his work was a ball that has 60 triangular faces, compared with about 20 in earlier Wilson balls; the triangles are formed by a series of intersecting "parting" lines to balance the asymmetry caused by the true molding parting line in all golf balls **(left).**

Every Ultra 500 ball has dimples of three different sizes, shapes and depths, each dimple mathematically positioned for best effect. The selection of dimples and their placement optimizes the interaction between opposing aerodynamic forces of lift and drag. Large dimples reduce air drag, enhance lift and maintain spin for distance. Small dimples prevent excessive lift that would destabilize ball flight to a degree. Medium sized dimples blend the characteristics of small and large dimples. The overall result, says Wilson, is a more uniform airflow over the spinning golf ball surface.

The research program that led to the patent-pending dimple pattern was conducted by Robert T. "Bob" Thurman, principal engineer, aerodynamics for Wilson Golf. Thurman's initial aerodynamic experience was acquired when he served in the Loads and Dynamics Group of Martin Marietta Manned Space Systems, New Orleans, Louisiana, manufacturer of the Space Shuttle's External Tank. Says Thurman:

"My responsibilities (at Wilson) include the design of dimple patterns for purposes related to the aerodynamic performance of golf balls, as well as analysis of trajectories and club/ball impact performance. In a way, my job is very similar to that at Martin Marietta. Instead of analyzing airloads on the Shuttle, I analyze airloads on airborne golf balls and the effects they have on the ball's trajectory, shape and distance. Instead of analyzing the slosh damping capability of the Shuttle's liquid oxygen tank, I analyze the effects of varying moments of inertia on the spin decay of a golf ball due to its liquid center."

*Ultra is a registered trademark of Wilson Sporting Goods Company.

VIDEO GOLF

Looking for an accurate way of measuring golf shots, engineer George Nauck conducted research in libraries and found that there were available a great many systems — cameras, sensors, computers — for analyzing the golf swing. But, he also found, the typical golf practice range provides virtually nothing in the way of measuring the result of a golf swing, the quality of a golfer's performance.

"As an engineer," he says, "I know something about the need for measurement and feedback to improve any process, and I felt that golf was no exception. There was clearly a need for down-range measurement on a practice range and for immediate accurate feedback to the golfer."

So Nauck filled the need by developing such a system himself, with a technological assist from NASA. Marketed by Nauck's company — known as ENCORE!!!, Jacksonville, Florida — it is called the ARPM (Advanced Range Performance Measurement) Video Golf System. An elevated camera located behind the tee follows the flight of the ball down range, catching the point of impact and subsequent roll. Instant replay of the video on a PC

monitor at the tee **(above)** allows measurement of the carry and roll. Distance and deviation from the target line are measured, as well as distance from a target when one is selected.

Groups of shots at a target are statistically scored to provide quality of performance information that serves either as an immediate basis for making adjustments or as a permanent record of skill level progress for golfers of all levels seeking to become a low handicapper. Such information is also important to the competitive professional golfer; tour pros have used the system to get immediate, precise information about carry and roll distances for every club in their bag.

In developing the system, Nauck found that library research was not enough for a determination of exactly what technologies were needed for a golf range measurement system. He contacted the Alabama Technology Assistance Partnership (ATAP), whose mission is to assist small businesses in identifying and accessing advanced technology. ATAP assisted him in preparing a problem state-

ment for submission to the Technology Transfer Office at Marshall Space Flight Center (MSFC). The request for assistance outlined Nauck's general concept and specifications for a golf range measurement system and asked what technologies might be applicable to such a system.

MSFC studied the concept, suggested video and image processing/computing technology, and also provided leads on commercial compa-

nies dealing in the pertinent technologies. Nauck **(shown above)** contracted with Applied Research Inc., Huntsville, Alabama for development of a prototype system, which was successfully field tested, debugged and refined in the summer of 1994. In October of that year, the ARPM system was introduced to the commercial market.

APPAREL MANUFACTURE

In pursuing its multiple responsibilities to the U.S. space program, Marshall Space Flight Center (MSFC) has of necessity developed new technologies over a broad spectrum of disciplines. Among these technologies is computer software for simulations, an area of particular importance to MSFC's primary responsibility of conducting R&D on space launch vehicles and methods of integrating vehicles and payloads. Adaptations of MSFC's simulation technology, now in wide use among America's clothing manufacturers, are enhancing the competitiveness of the U.S. apparel industry.

This technology transfer had its beginnings in 1989, when MSFC teamed with the University of Alabama in Huntsville (UAH) on a program involving development of advanced simulation software. In a separate action that occurred at about the same time, UAH was chartered by the State of Alabama's Department of Economic and Community Affairs to conduct a technology advancement program in support of the state's apparel manufacturers.

As UAH researchers worked on the two projects, engineers noted that certain major problems of the apparel industry might be solved by adapting the MSFC simulation software. For example, one prime need was a "try before you buy" computer simulation system that could analyze various apparel production layouts before implementing them on the sewing floor. A proposed production module could be pre-tested to determine its effectiveness; numerous options could be simulated on a "what if" basis to achieve optimum efficiency and productivity.

The possibility of adapting the NASA software to unit production systems and modular manufacturing systems was tested in several Alabama plants and found to be feasible. Therefore, in June 1992, MSFC awarded UAH a contract to develop an apparel-specific software package, one that would allow manufacturers to design and analyze modules without making an actual investment, and one that would work in ordinary PC equipment without the expense of additional software or specialized operator training. Completed and first made available to industry in mid-1993, the package includes two diskettes containing three different programs, along with instruction manuals and articles by UAH engineers on the implementation of modular manufacturing and its simulation in the apparel industry. Since then, MSFC has responded to requests for the package from more than 400 companies in 36 states; some of them have subsequently reported savings up to $2 million.

program usage. The MSFC visitors returned to NGC to teach the program to other company employees. The company has since extended the educational effort to include its engineering staff and all plant managers and supervisors. **At far left,** Larry Bannister, director of the engineering department, instructs costing engineer Donna Pilkingham on the use of the software.

NGC uses a modular manufacturing process in which modular "action teams" perform the cutting, sewing, finishing, tagging and packing; teams receive bonuses for achieving target production rates. The MSFC/UAH software is used to design and balance a modular line before committing to expensive hardware and logistics costs; to set up sewing lines; to predict production expectancy for an action team; to determine the composition of a new team; to

A typical user is National Garment Company (NGC), St. Louis, Missouri, which specializes in children's apparel. Late in 1993, while engaged in a switchover to modular manufacturing, NGC learned of the MSFC/UAH technology, obtained the package, and sent representatives to MSFC to learn further details of

vary the makeup of a team to allow adjustment of goals; and to determine the amount of time it takes to move a worker from one job to another.

Says Larry Martin, NGC vice president-manufacturing:

"We are impressed by the program's fairly simple operation, the amount of information we can apply to real-time situations, and the accurate and logical team working results."

MEMORY METALS

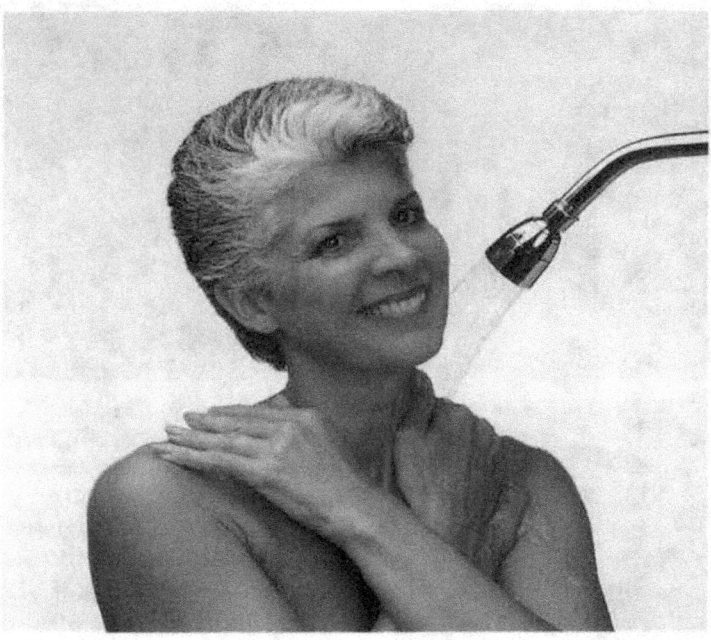

Shape memory effect, or SME, is a term for the ability of certain metal alloys to change from one shape to another in response to temperature changes, a direct result of a transformation of the alloy's crystal structure. An SME device can be made to expand when cooled or contract when heated; it may be a one-way or a two-way "memory." A one-way SME alloy can be deformed, then recover to retain permanently its original shape when heated to a certain temperature. A two-way alloy holds its original shape at one temperature and takes on another shape at a different temperature.

NASA explored this technology in the 1960s, shelved it for a time, then renewed interest in the 1980s, when the agency began preparatory work on the planned Earth-orbiting space station. Among companies awarded contracts for advanced SME investigations was Memry Technologies Inc., a subsidiary of Memry Corporation, Brookfield, Connecticut.

Under its NASA contracts, Memry produced alloys over a wide range of transformation temperatures in sheet, wire, rod and tube forms. The company developed several types of one-way memory quick connect/disconnect joints for space station structures and two-way memory actuators for the disconnect feature. In the course of this work, the commercial potential of SME technology became apparent and, in the latter 1980s and early 1990s, Memry adapted its NASA-acquired shape memory expertise to the commercial market and introduced a line of home and industrial safety products. Memry has since refined and advanced the technology, and developed some additions to the product line.

Among the current products are a trio of MemrySafe® units designed to prevent scalding from tap or shower water in the kitchen or bathroom. The heart of each system is a small valve that reacts to temperature, not pressure. When the unit senses dangerous temperatures (above 120 degrees Fahrenheit), it immediately reduces the water flow to a trickle. When the scalding temperature subsides, the unit automatically restores normal flow. This product line includes the ShowerGard® **(above,** with the simple valve insertion shown **at right);** the BathGard® for the tub spout **(above right);** and the Flow-Gard® **(far right),** which is installed easily in a bathroom, kitchen or utility room sink just like a standard aerator.

Memry products also include the FIRECHEK 2 and FIRECHEK 4 heat-activated pneumatic shutoff valves for industrial process lines. The FIRECHEK SME element, which needs no outside power source, senses exces-

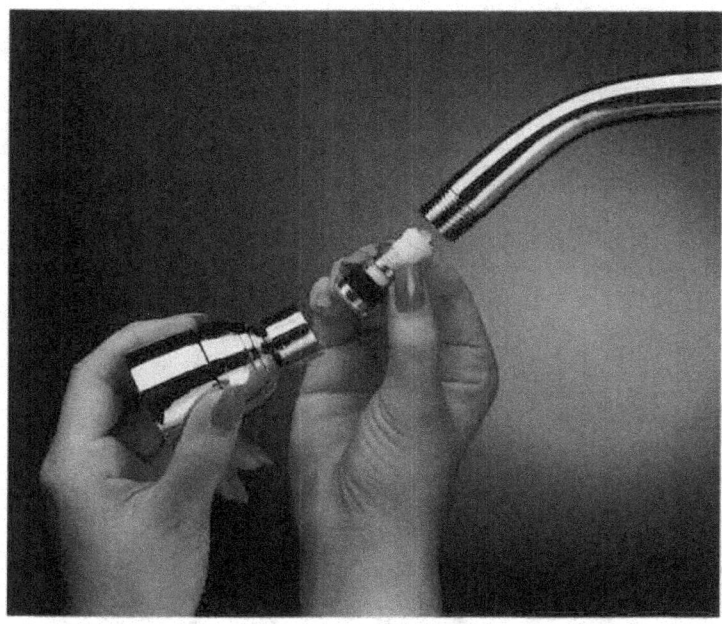

sive heat and automatically cuts off the pneumatic pressure that controls gas cylinders and process line valves.

The most recent product to emerge as a spinoff of the NASA technology is Memry's Demand Management Water Heater for residential heating, which employs a shape memory-activated diverter valve. Water is stored at 180 degrees in a special tank that serves as a heat exchanger. Incoming cold water is preheated at the heat exchanger, then delivered to the residential system at 135-140 degrees. When the water temperature in the tank drops below 140 degrees, the diverter valve bypasses the heat exchanger. This system shifts the electricity requirement from peak to off-peak demand periods. The consumer pays off-peak electricity prices for home hot water heating at significant savings, enough, Memry says, to recoup the cost of the system in a year and a half. Utility companies benefit from electrical load shifting through reduced need for construction of additional power generation plants.

™MemrySafe, ShowerGard, BathGard and FlowGard are registered trademarks of Memry Corporation.

WORKOUT MACHINE

The gyroscope-like device being demonstrated **at right** is the Orbotron, a tri-axle exercise machine patterned after a NASA training simulator for astronaut orientation in the microgravity environment of space. It is marketed by Orbotron, Inc., Carlsbad, California; more than 1,000 units are in service at health clubs, theme parks and other entertainment facilities.

The device was invented by Chris Altare, president of Orbotron, after he was able to study the NASA system. Altare was not interested in a weightlessness simulator as such; his primary interest was developing an exercise machine that would provide maximal physical benefits with minimal effort and eliminate the boredom associated with some types of exercise. "The Orbotron creates a sensation of weightlessness as does the NASA system," Altare says, "but the Orbotron is really a human-powered workout machine, unlike the NASA air-powered system."

A typical Orbotron workout takes only three to five minutes. The device has three orbiting rings corresponding to roll, pitch and yaw. The user is in the middle of the inner ring, support-ed by hand grips and foot restraints. Human power starts the rings spinning and the user can move in any direction. The mid-region of the stomach remains in the center of all axes; this, says Orbotron, eliminates any feeling of dizziness.

Orbotron provides a "full isokinetic and isometric workout" designed to tone and firm all major muscle groups, in particular the abdominal muscles. Company literature describes the operation of the device:

the user to receive up to $1\frac{1}{2}$ times his body weight plus $1\frac{1}{2}$ times the mass of the rings. Retoning and tightening the internal walls of the stomach is achieved, causing the stomach cavity to shrink."

A study of the Orbotron by the Department of Exercise Science, University of Southern California, reported that all of the 12 volunteer subjects (six male, six female) improved in all strength categories after eight weeks of five to seven minute workouts. The report

"The mass of the three rings at static rest is approximately 800 pounds, plus the weight of the human body. As the machine moves, dynamic forces of orbital motion cause a consistent change of the static and dynamic weight of the person inside the rings. It causes

concluded that "the Orbotron will improve aerobic capacity (cardiorespiratory fitness level). In addition, one can expect to find substantial improvement in both strength and endurance."

SCUBA WEIGHTS

Introduced in 1995, the Attitude Adjuster® pictured is a system for weight repositioning to adjust to a SCUBA diver's changing attitudes. A substitute for the conventional weight belt, the Attitude Adjuster is manufactured by Think Tank Technologies, Inc. (TTT), Stafford, Texas. The system was developed by TTT with the assistance of NASA's Mid-Continent Technology Transfer Center (see page 140) and Marshall Space Flight Center.

The Attitude Adjuster borrows a principle of submarine design technology — repositioning ballast/weight. Instead of the weight belt, compact tubes on each side of the diver's air tank permit controlled movement of lead balls within the Adjuster, automatically repositioning whenever the diver changes position. Micro adjustments for attitude stabilization are obtained by momentarily holding a desired position. The Adjuster has an emergency jettison system.

The advantages, according to TTT, are optimal weighting and ideal trim, thus reducing drag and energy requirements, and a new level of diver comfort made possible by elimination of heavy weights around the waist. Additionally, TTT says, the product contributes to lower air consumption, resulting in more bottom time.

The Mid-Continent Technology Transfer Center (MCTTC), located in College Station, Texas, is one of six NASA regional user assistance centers whose job is to provide information retrieval services and technical help to industrial and government clients. MCTTC's Texas Engineering Extension Service assisted entrepreneur Joseph J. Nicklo by performing market analysis; contributing technical help in the product design phase; helping to locate sources of capital; assisting in the development of a business plan; and arranging for testing of the Attitude Adjuster at Marshall Space Flight Center. NASA divers conducted the tests at Marshall's Weightlessness Environmental Training Facility for astronauts.

®Attitude Adjuster is a registered trademark of Think Tank Technologies, Inc.

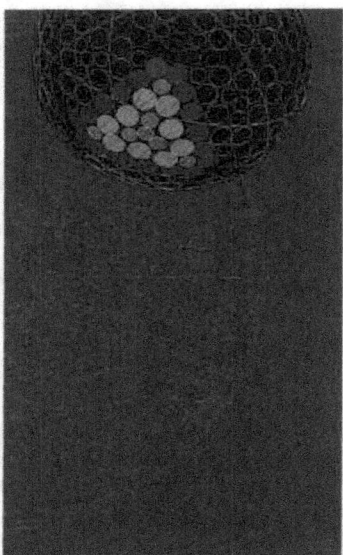

SKI BOOTS

Comfort Products, Ltd., Aspen, Colorado, a design and development firm that specializes in footwear, has been using NASA technology in its products for two decades. The NASA association began back in the 1970s, when Comfort Products adapted astronaut protective clothing technology to ski boot design. Specifically, the company borrowed heating element circuitry that kept Apollo astronauts warm or cool in the temperature extremes of the Moon, and used it to create built-in rechargeable footwarming devices that were supplied to leading ski boot manufacturers.

Comfort Products has once again applied NASA technology in a ski boot innovation, one that has proved highly successful with recreational skiers and has found favor among hundreds of world class competition skiers. In cooperation with Raichle Molitor USA, Inc. and the international Raichle organization headquartered in Switzerland, Comfort Products developed the Raichle Flexon concept in ski boots. The Flexon concept is an adaptation of the accordion-like corrugations of an extravehicular space suit joint, which offer astronaut flexibility of movement yet prevent kinking or distortion that might interfere with the suit's internal pressurization/temperature control equipment.

Raichle uses the Flexon technology in a score of different models of its extensive line of ski and snowboarding boots. The design features a convoluted, or corrugated, configuration of the tongue (front of the boot); this design permits the complex curved stiff plastic tongue to flex without substantial distortion. The Flexon

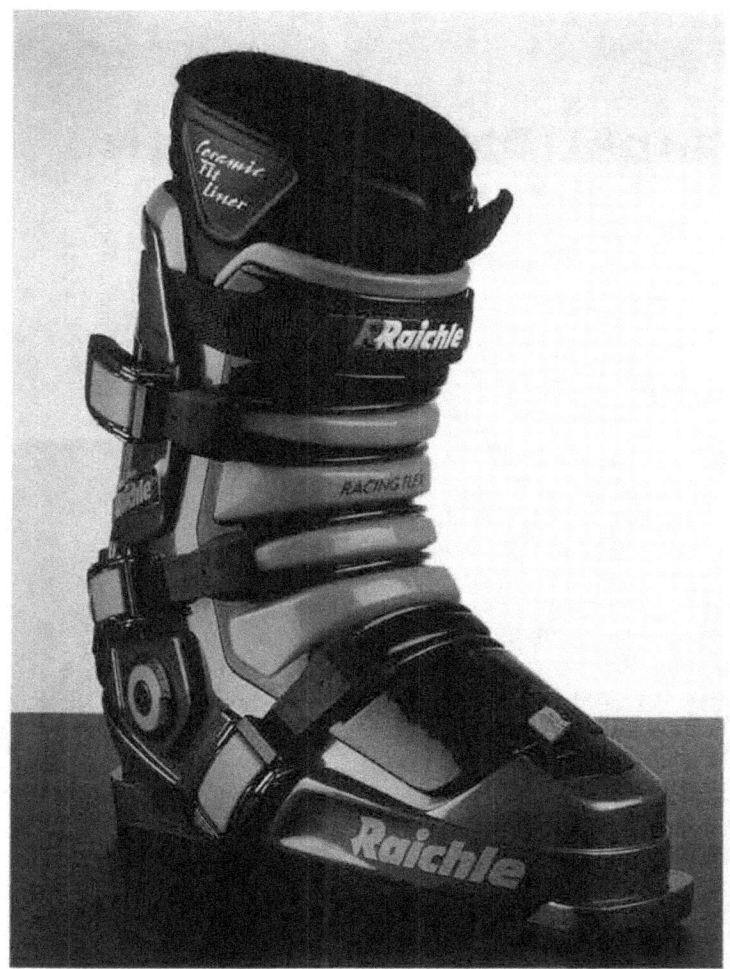

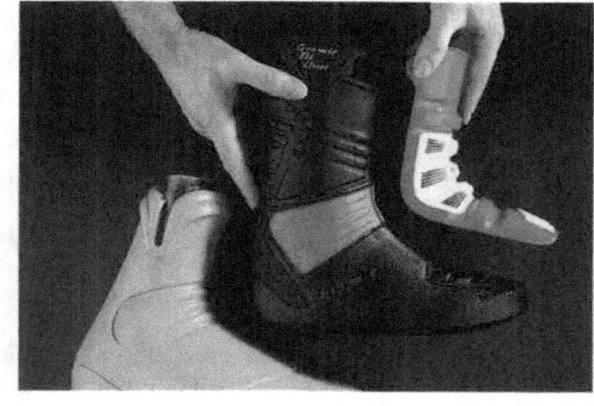

"optimized geometry," says Raichle sales literature, "provides immediate transmission of power and therefore precision skiing."

Says Erik O. Giese **(left)** of Comfort Products: "The idea came from the joints in the space suit, which required articulation without distortion of a pressure vessel. It is also similar to a vacuum cleaner hose flexing without destruction, versus a paper towel tube that will bend and crimp when flexed."

The technologically-advanced Raichle line also offers a "ceramic fit" system that assures a perfect fit for every individual foot. The system consists of a chambered "breathing" bladder, filled with tiny hollow ceramic spheres, sewn into the innerboot. When the user puts a foot into the boot, the spheres adapt to the contour of the individual foot. When the boot is buckled, the air is pressed out of the bladder and the spheres can no longer move. This, says Raichle, provides a precise "hold" for optimal transmission of power.

Above, Erik Giese displays a boot last, or mold (green), the bladder (black and maroon) and the tongue of the boot (red and yellow). The photo **at left** illustrates the lineage of Comfort Products' boots: on the left is the newest model, in the middle (red) a model with the Flexon technology, and at right a model without the ribbed tongue. The newest model is shown in closeup **at left above.**

RADIATION INSULATION

The Apollo spacecraft was a masterpiece of temperature control technology. It had to be, to protect astronauts from temperatures that ranged 400 degrees above and below zero Fahrenheit.

A prime element of the environmental control system that permitted astronauts to work inside the Command Module in shirtsleeves was a highly effective radiation barrier. Made of aluminized polymer film, it barred or let in heat to maintain a consistent temperature in an environment where ordinary insulative methods would not have sufficed. The key was aluminization of the material, which provided a reflective surface that kept more than 95 percent of the radiated energy from reaching the spacecraft interior.

The radiation barrier has since been used on virtually all spacecraft, including unmanned spacecraft where delicate instruments need protection from extremes of temperature. It is also a key element of a sophisticated energy conservation technique for home and office building installations, plus a variety of special applications.

An example of companies supplying variations of the space-developed material is Quantum International Corporation, Puyallup, Washington. Quantum markets several tri-laminated insulating materials known as Super "Q"® Radiant Barrier. Quantum's insulation is a combination of high grade industrial aluminum foil overlaid around a central core of another material, usually fire-resistant propylene or mylar. This central layer creates a thermal break in the structure, thereby allowing low values of radiant energy emission. The highly polished surface of the aluminum reflects as much as 97 percent of all infrared radiant energy (heat).

Quantum supplies materials for domestic and international contractors who use Super "Q" Radiant Barrier for thermal insulation of homes and office buildings, and for a variety of industrial applications. A novel application for housing units is displayed **at left** by Quantum's senior vice president, Preston E. Smith, and Darrel A. Duisen, president of Global Development and Trading Corporation, Pauma Valley, California, which manufacturers these modular panels in the U.S. and abroad. **Above** is a cutaway panel, showing the Super "Q" insulation sandwiched between an expanded polystyrene core strengthened with a steel wire grid. The panel weighs less than 30 pounds and can be lifted to heights by hand or winch pulley, which gives it special utility in undeveloped countries where cranes are often not available. The panels can be used for walls, floors or ceilings.

In addition to serving the traditional home and indus-

trial market, Quantum is supplying Radiant Barrier for mobile applications, such as autos, trucks and food transports. **Above,** Quantum International president Mike Kerschner (foreground) displays a Quantum Cool Wall® panel while technicians of Cryo-Power Inc., Astoria, Oregon complete installation of Cool Wall in a cryogenic mobile container. The lightweight panels take up less space while providing superior insulation value. Grocery and produce transporters use Super "Q" panels to line their trailers and containers, warding off unwanted heat during transportation to reduce spoilage and costs.

Auto and truck manufacturers are also using Super "Q" Radiant Barrier as reflective thermal liners in certain parts of the vehicles to maintain passenger comfort by dampening sound and by blocking engine, exhaust and solar heat. **At left** are Garth Moore and Ken Loomis (hand on hood), co-owners of U-B-Kool, Reno, Nevada, a company that installs Super "Q" products in autos and trucks. The closeup **above** shows the under-the-hood insulation.

®Super "Q" and Cool Wall are registered trademarks of Quantum International Corporation.

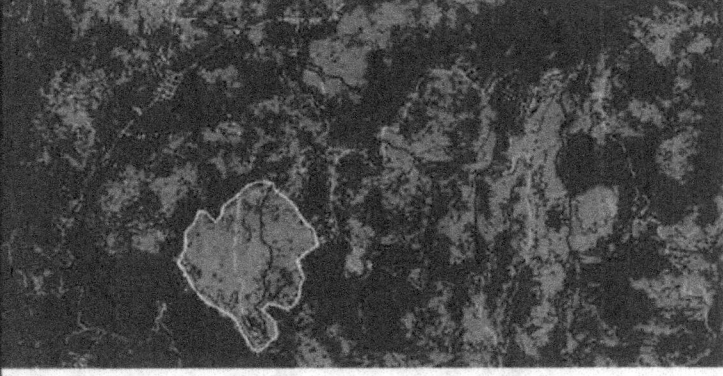

Among environment-related technology transfers is a NASA/private sector partnership effort to expand commercial use of remotely sensed data

Remote sensing technology offers a broad spectrum of opportunities that could lead to new commercial products or processes of benefit to the nation as a whole and to individual companies. However, although sales of remote sensing data are on the rise, the government is still the principal market for such data; industry has been slow to grasp the potential.

In an effort to accelerate the growth of commercial remote sensing in the national interest, NASA has created a number of mechanisms designed to promote greater awareness among industry firms of the remote sensing potential and to provide basic training in the commercial application of the technology. Among the mechanisms is the Visiting Investigator Program (VIP), managed by Stennis Space Center.

VIP is oriented toward companies that might be able to integrate remote sensing into their operations to create new products and new markets, or to improve existing product lines. Recognizing that such companies generally have limited resources to invest in exploring new technologies, VIP offers a chance for a firm to expand its capabilities on a low-risk basis; VIP costs them nothing but their own employees' travel expenses.

A prospective VIP participant submits a proposal involving use of remote sensing in an actual project, detailing research objectives and outlining the commercial possibilities. If the proposal is accepted, Stennis Space Center acquires the necessary remote sensing data, instructs company personnel in the use of remote sensing hardware and software, and provides guidance in executing the project. A typical project runs six to nine months; if it produces promising concepts, the potential exists for further NASA/company joint research.

An example of a VIP project is the experience of Community Coffee, Inc., Baton Rouge, Louisiana, an importer buying coffee primarily from Guatemala and Brazil. To maintain competitiveness, the company needs advance information about a growing area's crop yield and quality at the next harvest. But the information coming out of producing countries is usually sketchy and sometimes questionable. Looking for an efficient way of monitoring coffee growth and yield trends, Community Coffee undertook a VIP project involving the use of remote sensing data and Geographic Information System (GIS) technology as a means of tracking coffee growth and assessing crop health.

Stennis Space Center provided satellite imagery of an area in Guatemala and supplemented it with infrared aerial photography of the same area. Working with NASA remote sensing experts, Community Coffee's George Guthrie, director of green coffee purchasing, enhanced the imagery in a multistep process, adding ground reconnaissance information about the production of individual coffee *fincas* and surrounding environmental conditions, together with such GIS data as the location of urban areas, roads, contours, hydrography, etc. By computer processing, the investigators eventually produced a composite image of the area showing (by color coding) the size and location of coffee-growing areas, along with a three-dimensional perspective of the same area.

At left is an enhanced satellite image of coffee growing areas in a region of Guatemala, developed in a multistep land classification process that ruled out non-coffee land (red) and highlighted coffee fincas (gray). **Below** is a three-dimensional perspective of the same area in which elevations and other topographic features have been added. Community Coffee, a Louisiana importer, used satellite imagery, aerial photography and geographical information in a prototype study that explored the potential of developing coffee crop yield forecasting maps.

These products represented an experimental prototype of what Community Coffee hopes will be an efficient system of forecasting coffee crops. Using satellite imagery, the company seeks to produce detailed maps that will separate coffee cropland from wild vegetation and show information on the health of specific crops. This information is economically important. Timely knowledge that the coffee crop yield in a given area would be greater than average could result in lower prices at harvest time; therefore, the company's buyers could delay purchases until prices drop. If satellite imagery indicates a lower than average crop, meaning higher prices, Community Coffee could turn to other coffee sources to hold prices down.

Community Coffee's interest in remote sensing goes beyond procurement strategy. The company is looking into the possibility of marketing this sophisticated forecasting technique to other coffee buyers, or to coffee farmers and government agricultural services, who could use the information to optimize application of fertilizers, pesticides and irrigation for maximal crop yields, minimal pollution and lower production costs.

(Continued)

Nesbit Environmental, a Louisiana firm, investigated the potential for locating abandoned barges in the Mississippi River and channels using airborne and spaceborne imagery. This sample image was developed by combining Landsat satellite data with Side-Looking Airborne Radar data.

Landsat TM/Radar Merge
12 meter resolution

Begun in 1988, the NASA Visiting Investigator Program (VIP) was designed to bring remote sensing applications closer to commercialization by conducting viability demonstrations toward creation of new products and processes. Industry investigators are granted access to Stennis Space Center's commercial remote sensing facilities and its sophisticated data processing hardware and software. Working with NASA scientists in all aspects of a project from planning through product development, partner companies become familiar with the theoretical and practical bases for remote sensing, airborne/spaceborne data acquisition, related instrumentation, and data analysis techniques.

While the primary aim of both NASA and the industry partner is commercialization of the technology, participation in the VIP program often leads to broader long term benefits. An example is the case of Scott Nesbit of Nesbit Environmental, Inc., Baton Rouge, Louisiana, whose VIP work led to a merger with a larger environmental consulting firm.

Nesbit's VIP project was an investigation of the potential of remote sensing for locating abandoned barges in the Louisiana coastal zone. Over the past 40 years, barges and other vessels have been abandoned in the channels and adjacent waterways of the Mississippi River, a major corridor for petrochemical transportation. These barges may become illegal disposal sites for used oil or hazardous materials, which may leak and cause serious pollution of the waterways and coastal estuaries.

However, locating abandoned barges is not easy. The vessels are scattered over a very wide area, seasonally covered by vegetation and undetectable; additionally, the job of differentiating active vessels from abandoned barges requires multiple observations showing the barges anchored in the same location over a long period.

Since 1989, the U.S. Coast Guard and the Environmental Protection Agency have been conducting a program to locate abandoned barges that represent environmental risks. Nesbit Environmental's VIP project was an investigation to determine the detection capability and cost effectiveness of using the U.S. Geological Survey's Side-Looking Airborne Radar (SLAR) together with the Landsat Thematic Mapper (TM), a satellite-based scanning instrument, and panchromatic data from the French SPOT remote sensing satellite. The effectiveness of these data sources was to be determined by comparing their digital remotely-sensed data with color infrared aerial photography.

Assisted by Stennis Space Center experts, Nesbit initiated a process whereby aerial photography was employed as the principal data source with the SLAR, Landsat TM and SPOT as

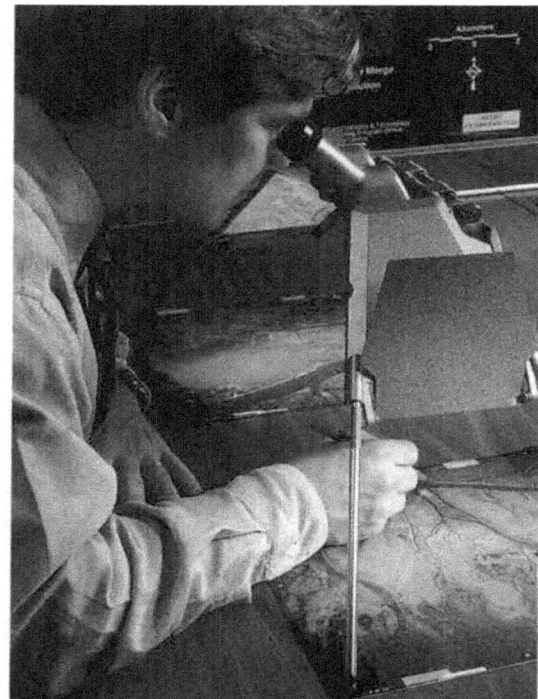

Scott Nesbit of Nesbit Environmental is viewing aerial images of the Louisiana bayou through a stereoscopic viewer that allows easier and more definitive identification of features.

A typical aerial image used in Nesbit's VIP project for locating abandoned barges in the waterways of the Mississippi River complex.

secondary sources for verification. To provide the abandonment time factor, SLAR and satellite imagery acquired at several times from 1988 to 1994 was used. Potential vessel locations were screen digitized to map coordinates for field checking.

The five-month investigation concluded that a combination of aerial photography backed by SLAR and satellite imagery offered a cost-effective way of identifying potentially abandoned vessels for follow-up field check by boat or plane. Although Landsat TM data were not very useful in initial barge detection (due to coarse resolution), the data were helpful when merged with SLAR data to provide information regarding land/water delineation. The advantages of combining color infrared aerial photography with SLAR data include increasing the overall detection capability by using an additional detection method, and providing multiple surveillance dates to verify abandonment cost-effectively (due to the ready availability and low cost of SLAR data). The combined system offers increased accuracy and reduced cost of follow-up inspections to determine the status of suspect vessels.

The VIP project not only achieved its primary objective and provided new technology for use in barge detection, it also attracted the attention of C-K Associates, Inc., also of Baton Rouge, and led to a merger of Nesbit Environmental and C-K. Where C-K Associates had previously limited remotely-sensed data acquisition to aerial photography, the merged company has since invested in Geographical Information System (GIS) hardware, software and technical training, and initiated the use of a GIS to process multiple sources of remotely-sensed data as an on-line engineering tool.

91

ENVIRONMENT
AND RESOURCES
MANAGEMENT

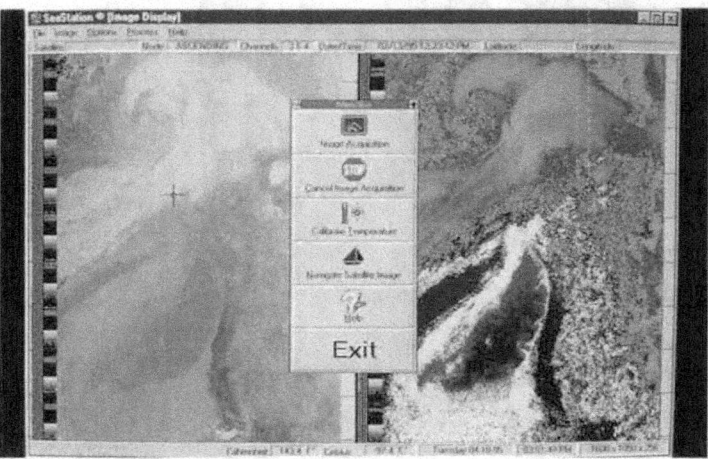

COMMERCIAL EARTH OBSERVATION

A close relative of the NASA VIP program (see page 88) is the government/industry cooperative program known as EOCAP (Earth Observation Commercial Applications Program). Designed to help private industry develop and commercialize applications

of remote sensing, EOCAP is an effort to encourage private investment in, and broader use of, NASA technology for gathering and analyzing information about Earth's land and ocean resources.

Like VIP, EOCAP is managed by Stennis Space Center, but the programs differ. VIP generally involves short-term projects of three to six months with NASA providing the funding; EOCAP projects can run three years or more and funding is shared by NASA and the private sector participant.

An EOCAP example is the development of a new tool for analyzing remotely sensed data, a process that provides a subpixel detection capability for increased accuracy and significantly enhanced discrimination among surface features in imagery.

Known as the Applied Analysis Spectral Analytical Process, or AASAP, the program was developed by

Applied Analysis, Inc. (AAI), Bellerica, Massachusetts. **At left** are AAI president Donald B. Damm, (seated) and the company's chief scientist, Dr. Robert L. Huguenin, who developed AASAP.

As the term implies, subpixel processing detects or classifies objects that are smaller than the size of a pixel. In most remote sensing applications, image pixels are "mixed pixels," meaning that they contain not only the unit of interest but also "background" features that distort the spectral properties of the unit of interest, resulting in a loss of discrimination and potential misclassification.

AASAP enhances the classification process by removing the background. To do so, the software must identify the background matter in each pixel, then remove exactly the right amount of it. The residual spectrum (what's left after background removal) is a relatively pure representation of the unit of interest. **At top right** is a comparison

 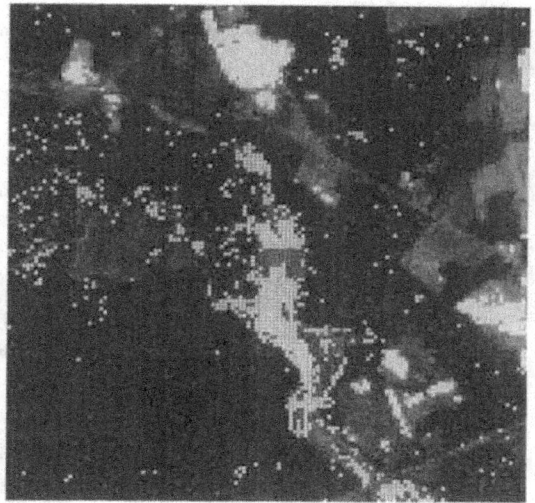

between normal satellite imagery (at left) and a subpixel processed image; the yellow/red area in the right image represents the additional information obtained by subpixel processing.

According to its developers, AASAP is achieving accuracies of about 90 percent. It is now being offered as a modular addition to the ERDAS IMAGINE software package marketed by ERDAS, Inc., Atlanta, Georgia for remote sensing applications.

Another EOCAP example involves the development of the SeaStation ocean information systems by Ocean and Coastal Environmental Sensing (OCENS), Seattle, Washington. Through an EOCAP partnership, OCENS developed SeaStation, a low-cost, portable, shipboard satellite groundstation integrated with vessel catch and product monitoring software.

SeaStation targets marine users in the commercial and recreational fishing markets, along with recreational boaters. It provides a Windows-based graphical interface that permits the user to acquire and process environmental data originating with satellite-based sensors. Acquisition and processing software is linked to the Global Positioning System data stream to provide real time relationships between vessel position and such data as sea surface temperature, weather conditions and ice edge location. This allows the user to increase fishing productivity and improve vessel safety. **At left,** OCENS president Mark H. Freeberg and a client, an officer (red clothes) of a commercial fishing vessel, check the output of the SeaStation system. On the **opposite** page, top, is a representative SeaStation screen showing ocean temperature gradients and ship position data.

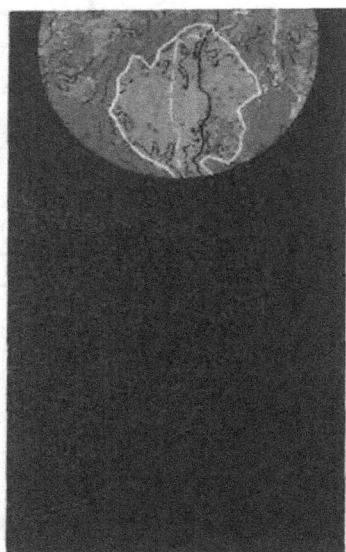

WEATHER INFORMATION SYSTEM

Weather is a critical factor in aircraft operations. It is the largest single contributor to flight delays and a major cause of aircraft accidents.

A study conducted for NASA by Ohio State University reported that the principal difficulties in making proper flight decisions are the timeliness and clarity of weather data dissemination. Properly equipped aircraft can get printed weather reports, but the most common means of getting weather information is by voice communication with ground sources. Word of mouth reporting over congested radio channels is at best inefficient and at times extremely difficult — for example, in critical weather when a large number of airmen are working the radio.

To advance the technology of in-flight weather reporting, Langley Research Center developed — in the early 1990s — a cockpit weather information system known as CWIN (Cockpit Weather Information). The system draws on several commercial data sensors to create radar maps of storms and lightning, together with reports of surface observations.

Shown above is a CWIN display in the simulation cockpit of Langley's Transport Systems Research Vehicle (TSRV), a modified jetliner used to test advanced technologies. The CWIN display is the lower right screen among the four center panel screens; in this instance it shows color-coded categorical weather at airports throughout the U.S. By pushing a button, the pilot may select from a menu of several displays, such as a ceiling/visibility map, radar storm map, or lightning strike map **(right above).**

In simulation tests during 1993, pilots from several airlines and aerospace manufacturing companies flew the CWIN-equipped TSRV. A consensus indicated that CWIN offered improved safety, better weather monitoring and substantial fuel savings, due to fewer miles detouring

around storms. Principal CWIN developer Charles H. Scanlon, Langley aerospace technologist, estimated that a typical U.S. airline could cut its operating costs by $5.9 million annually with a system like CWIN.

This NASA technology has been adapted to commercial use by ARNAV Systems, Inc., Puyallup, Washington. Now in service aboard general aviation aircraft is the company's System 6, a key feature of which is a weather datalink known as WxLink™. ARNAV's WxLink employs several CWIN features, among them are the lightning graphics and ground-based radar summaries.

WxLink is an aviation weather system based on advanced airborne sensors, precise positioning available from the satellite-based Global Positioning System (GPS), cockpit graphics and a low-

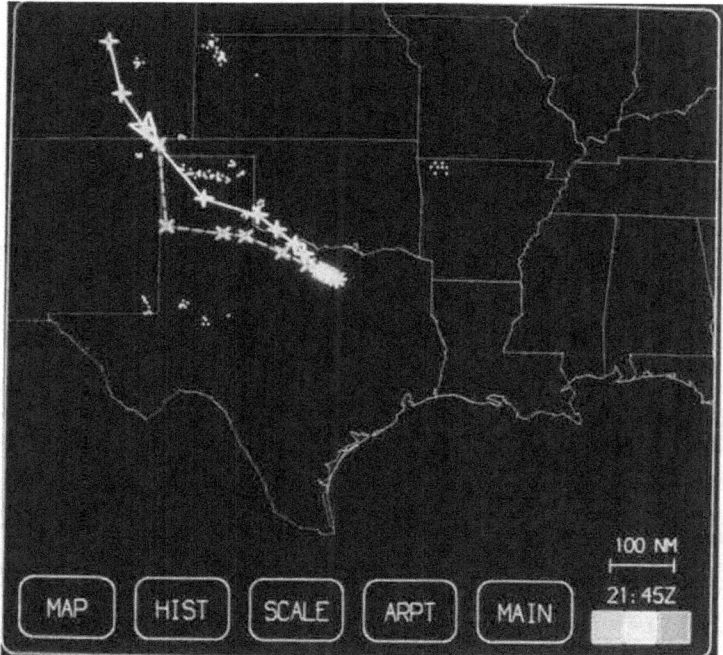

cost datalink. It is a two-way system that uplinks weather information to the aircraft and downlinks automatic pilot reports of weather conditions aloft.

PanAm Weather Services, Minneapolis, Minnesota, which operates 312 Weather-Mation sites in 38 states, provides the weather data input, regularly updated, for the uplink from ground to cockpit. ARNAV airborne equipment transmits PIREPS weather reports via satellite back into the National Weather System for dissemination to other users.

ARNAV's System 6 is not one product but a family of products designed for integration into a complete application. The cockpit display includes, in addition to the weather information, moving map navigation, engine and environment monitoring, a checklist library, condition alerts and digital messaging. The key unit is the MFD Multi-Function Display (MFD 5000 in monochrome, MFD 5100 in color), which packs the performance of several systems into a single display.

At right, the MFD 5100 is shown on the instrument panel of a demonstrator light twin; **below** is a closeup of the weather and other informational displays available to the pilot. The MFD performs health checks on 70 onboard systems and alerts the crew to difficulties; it is fully integrated to deliver crew advisories on navigation, fuel management, environment, air data, engine and airframe condition; and it watches for off-altitude and altitude clearance situations.

™WxLink is a trademark of ARNAV Systems, Inc.

CONTOUR MAPPING

In 1989-90, the Ohio State University Center for Mapping (CFM), one of NASA's Centers for the Commercial Development of Space, developed a vehicle known as the GPSVan System to meet a need for a cost-effective way of creating accurate highway maps from digitized data.

The "GPS" stands for Global Positioning System, a constellation of satellites operated by the Department of Defense to provide highly precise position information for aircraft, surface vessels and ground vehicles. The GPSVan is a mobile data collection system for mapping roads and their surroundings while driving at regular highway speeds; GPS receivers acquire satellite signals, which are computer processed on board the van, for a continuous record of the vehicle's location.

The van also has two video cameras, which scan the local terrain and produce images whose location is coordinated with the van's position at the time of imaging. Features such as street lights, utility poles, manholes and roadside structures can thus be positioned on the base maps. All this data is fed into a computerized Geographic Information System (GIS). The expense of switching map data to electronic databases is eliminated because the digitizing has already been done aboard the van.

The GPSVan has been used in commercial mapping operations but CFM continues to advance the technology. A novel and commercially promising advancement is the use of satellite positioning technology to determine the contours of a construction site. That was accomplished in a three-year project jointly conducted by CFM and George J. Igel & Company, Inc., Columbus, Ohio.

In 1994, Igel was able to use the GPS survey system to determine the contours of a construction site by driving over it in regular patterns. Using GPS receivers and

software produced by CFM, Igel employed an antenna-equipped pickup truck to collect data on ground elevations, which were recorded and registered as contours. The Igel/CFM team was able to cover about six acres per hour with complete contours plotted "on the fly." The data proved accurate to within one-tenth of a foot in all dimensions.

At left is a view of a vehicle based system at a construction site. A closeup of the base station receiver and antenna is shown **at right.** **Below** is a view of the vehicle-mounted antenna that receives the satellite signals, which are used to calculate the coordinates of points on the ground.

AMT Systems Engineering, one of CFM's industrial partners, designed the enclosure for the survey system; a more advanced final product has been configured for a

backpack instead of the truck. The GPS survey system will be used to reduce the time required for mapping and staking, and to monitor the amount of soil moved on a given day.

Igel and CFM have continued development work on the construction application toward an eventual capability to obtain elevations while using earth-moving equipment. The team has conducted experiments in which GPS positioning is used to control a cutting blade on earth moving equipment for total site grading by means of digitized design information.

In these experiments, a computer and TV-like monitor display the digitized site design to the machine operator. Using GPS, the machine's position on the site is deter-

mined and communicated via the computer screen to the operator. By comparing the machine's position to the final design, the operator determines how much cutting and filling is required. A graphical user interface is being designed to assist the operator in evaluating the site and to support on-the-fly decision making. Igel and CFM envision a future system capable of feeding GPS signals and design information directly to the vehicle's hydraulic controls to automate the earth cutting and filling function.

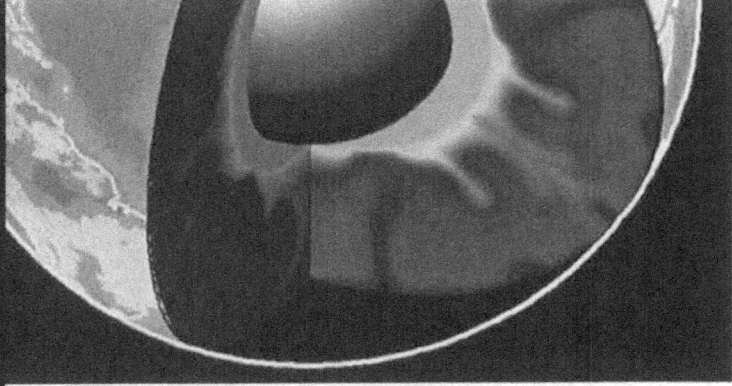

A Space Shuttle-derived information management system with broad potential for private industry applications leads a selection of spinoffs in the field of computer technology

At current schedules, each of NASA's four Space Shuttle Orbiters must fly two or three times a year. On the surface, that might not seem much of a problem. But a little known part of the Space Shuttle story is the incredibly complex process of preparing an Orbiter for its next mission.

Much of the work is accomplished in the Orbiter Processing Facility (OPF) at Kennedy Space Center. There technicians must repair the heat shielding tiles and other areas of the Orbiter's exterior damaged by reentry into the atmosphere; they must repair or replace on-board equipment that caused in-flight anomalies on the previous mission; payloads from the last mission must be removed, a task much more complex than it sounds; new payloads must be installed and integrated, demanding many modifications to the Orbiter; the crew module must be reconfigured for the new mission; and, in the final phase, every component of every system in the Orbiter must be thoroughly tested for individual function and complete system integrity.

All of that is only a skeletal summary of the demanding, intricate work that goes on at the OPF. One can get a general idea of the extraordinary effort involved by these statistics: the average "flow" — the complete cycle of refurbishing an Orbiter — requires the integration of approximately 10,000 work events, takes 65 days and some 40,000 technician labor hours.

Under the best of conditions, scheduling each of the 10,000 work events in a single flow would be a task of monumental proportions. But the job is further complicated by the fact that only half the work is standard and predictable; the other half is composed of problem generated tasks and jobs specific to the next mission, which creates a highly dynamic processing environment and requires frequent rescheduling.

For all these difficulties, Kennedy Space Center and its prime contractor for Shuttle processing — Lockheed Space Operations Company (LSOC) — are doing an outstanding job of managing OPF operations. They have been helped in that regard by a number of processing innovations in recent years. One of the most important is the Ground Processing Scheduling System, or GPSS. The GPSS is a software system for enhancing efficiency by providing an automated scheduling tool that predicts conflicts between scheduled tasks, helps human schedulers resolve those conflicts, and searches for near-optimal schedules.

This unique system has helped streamline OPF scheduling. It also offers industrial benefits of substantial order and a commercial version of the software is now available to private industry through a newly-formed company known as Red Pepper Software Company, San Mateo, California.

GPSS is a cooperative development of Ames Research Center, Kennedy Space Center, LSOC and a related company, Lockheed Missiles and Space Company. It originated at Ames, where a group of computer scientists conducted basic research on the use of artificial intelligence techniques to automate the scheduling process. A product of that work was a software system for complex, multifaceted operations known as the Gerry scheduling engine.

Kennedy Space Center brought Ames and Lockheed together and the group formed an inter-center/NASA contractor partnership to transfer the technology of the Gerry scheduling

At left, Kennedy Space Center technicians are preparing a Space Shuttle Orbiter for its next mission, an intricate task that requires scheduling 10,000 separate events over 65 days. A NASA-developed computer program automated the scheduling job and spawned a commercial spinoff, demonstrated *(below)* by Monte Zweben, president of Red Pepper Software Company.

engine to the Space Shuttle program. The transfer was successfully accomplished and GPSS has become the accepted general purpose scheduling tool for OPF operations.

The Space Shuttle Ground Processing Project was managed for NASA by Monte Zweben, who had designed and developed several planning and scheduling systems while serving as a deputy branch chief at Ames Research Center. In January 1993, Zweben left Ames and, with finance expert Daniel T. Doles, founded Red Pepper Software Company (RPS). In August of that year, Red Pepper received a license from NASA to commercialize the GPSS system.

The RPS software systems derived from the Shuttle scheduling technology are called ResponseAgent™ products. ResponseAgents, company literature says, "are a new kind of software, providing users with an intelligent assistant to constantly monitor manufacturing variables, report issues and develop optimized solutions to complex problems. The products start with the fundamental objective of satisfying customer demand, then optimize materials, capacity and labor in real time to meet the objective." ResponseAgents are designed to complement existing transactional and shop control systems.

Red Pepper's initial product line includes three systems: the Enterprise ResponseAgent, which is designed to integrate the worldwide factories and distribution centers of large companies into one cohesive manufacturing enterprise; the Production ResponseAgent, offering integrated planning and scheduling for production facilities; and the Distribution ResponseAgent, which does the same for distribution centers.

RPS emphasizes real-time responsiveness in the modern manufacturing environment as a primary benefit of its products. Says the company: "In the past, manufacturers could develop a monthly plan and execute. Today, the market requires the flexibility and responsiveness to deal with real-time unpredictability while still satisfying customer requirements. This level of responsiveness requires a new approach to planning and scheduling, one that responds proactively to real-time change. ResponseAgents provide this new approach."

™ResponseAgent is a trademark of Red Pepper Software Company.

ASTRONOMY AID

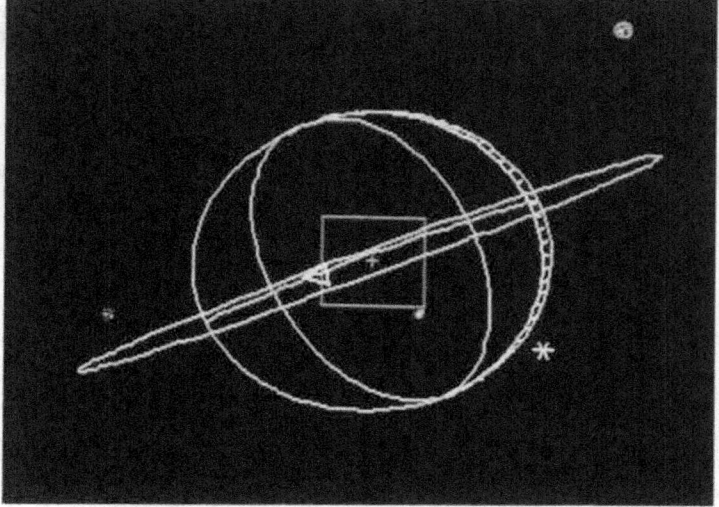

In 1977, NASA launched the Voyager 1 and 2 space probes on an indefinite duration odyssey that included flyby encounters with the outer planets Jupiter, Saturn, Uranus and Neptune. The two spacecraft continued past Neptune on diverging paths that will eventually take them out of the solar system; for the interim, they are conducting the Voyager Interstellar Mission, reporting solar wind/solar cycle data and trying to determine the location of the heliopause, the boundary that represents the outer edge of the Sun's magnetic influence and the beginning of interstellar space.

In the mid-1980s, Jet Propulsion Laboratory (JPL) developed a computer program called MIP (Multimission Interactive Planner) to help astronomers analyze scientific and optical data collected on the Voyager's Grand Tour. MIP was also intended as an analysis tool for such other space missions as the Giotto spacecraft's 1986 rendezvous with Comet Halley: the Hubble Space Telescope's views of the universe; and the Galileo encounter in Jupiter, which begins in December 1995. A commercial version of the program called XonVu, pub-

lished by XonTech, Inc., Van Nuys, California, was created by John D. Callahan of Sun Valley, California, who had designed the original MIP while serving as an astronomer at JPL.

Callahan has since developed two more advanced programs based on MIP technology and made them commercially available. One, known as Grand Tour, has appeared in a number of software catalogs, including Andromeda Software, Inc., Amherst, New York; the other, Jovian Traveler, was published by Zephyr Services, Pittsburgh, Pennsylvania. The programs include refinements to MIP and improvements over XonVu, including pop-up menus that appear simultaneously with the graphic display.

The software packages simulate the missions of Voyager 1 at Jupiter and Saturn, Voyager 2 at Jupiter, Saturn, Uranus and Neptune, and Giotto in close encounter with Comet Halley. **Above** is

a simulation of Jupiter, its ring, and three of its moons as viewed by Voyager 1 on its 1979 approach to the planet; the green squares represent the fields of Voyager's wide angle and narrow angle cameras.

With these programs, space novices and professional astronomers can generate scenes of the planets, moons, stars or other objects seen by the Voyagers, or Comet Halley's nucleus and tail as viewed by Giotto. The programs take the user along the actual paths flown by the Voyagers and Giotto; a user cannot invent a new flight path. He can, however, zoom in on an object, rotate the field of vision, or change perspective. For example, he can view a space scene from the perspective of the spacecraft, from Earth, or from a point in space that provides a view of both the spacecraft and the encounter target. The programs can be used on a wide range of PCs.

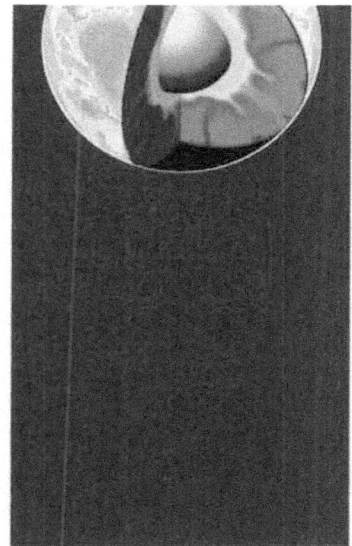

IMAGING SYSTEM

At right is the DTI 1100C Virtual Window™, an advanced three-dimensional imaging device developed by Dimension Technologies, Inc. (DTI), Rochester, New York. The system won a 1994 R&D 100 Award presented by R&D Magazine, which annually recognizes the 100 most technologically significant new products of the year. The 1100C Virtual Window is based on technology developed under NASA contracts.

The Virtual Window is termed by its developers "a major breakthrough in flat panel display technology, one that gives true 3D imagery without special glasses." Commercial 3D systems are grouped in two broad categories: stereoscopic, in which the user wears glasses or employs an optical device to perceive a 3D image, and autostereoscopic, in which 3D images can be viewed without optical aids. DTI has developed a number of autostereoscopic flat screen displays, which offer advantages in many applications.

The DTI Virtual Window employs an innovative illumination system to deliver the depth and color of true 3D imaging. The system can provide enhanced viewing for critical endoscopic surgery procedures, Magnetic Resonance Imaging scans, or surgical planning. In hazardous area applications, it offers optimal viewing of teleoperated robots for waste removal, reconnaissance, remote inspection and deep hole mining. It can improve the reality of simulators for training, and it has utility in aviation as an in-cockpit display to improve pilot "situation awareness" in all types of aircraft.

ing. Introduced in 1993, Virtual Window is a spinoff from technology developed under NASA Small Business Innovation Research (SBIR) grants and other NASA contracts awarded by Ames Research Center, in particular a contract for development of "a large autostereoscopic display for scientific visualization applications."

™Virtual Window is a trademark of Dimension Technologies, Inc.

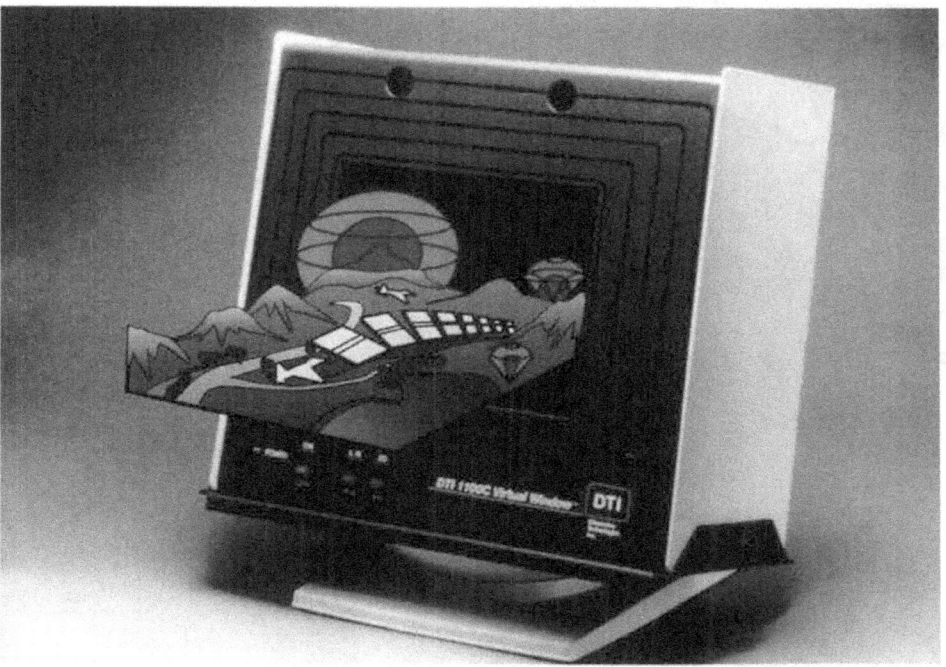

DTI is a relatively new company, founded in 1986 to pursue autostereoscopic display technology. In 1993, under Air Force contract, DTI developed a cockpit stereoscopic system for night flying.

101

COMPUTER
TECHNOLOGY

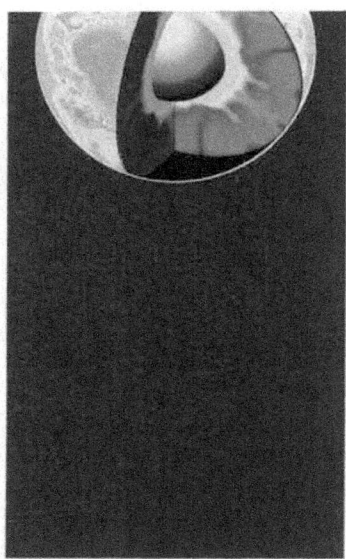

SOFTWARE BRIDGE

I-Kinetics™, Inc., Burlington, Massachusetts, is a new software company formed to develop high performance distributed computing products that integrate applications across multiple operating systems and frameworks. The company focuses on developing practical applications solutions to be deployed in large distributed information systems that integrate UNIX relational databases and UNIX applications with popular Microsoft Windows applications. I-Kinetics' products are used to support securities trading, financial management, operations, planning and scheduling, integrated design and manufacturing, and project management applications.

Among the company's products is the I-Bridge® family of software systems that allows users of Windows applications to gain quick, easy access to databases, programs and files on UNIX services. Information goes directly onto spreadsheets and other applications; users need not manually locate, transfer and convert data.

I-Bridge is a commercial version of core software developed under a Small Business Innovation Research (SBIR) grant from NASA, under which the company developed and refined the Object Express data encoding/decoding technology that makes it easier for Windows applications on the PC to "talk" with UNIX services.

I-Bridge is in commercial service with a variety of customers. Example: in a major mutual fund investment company, fund portfolio managers are using I-Bridge regularly to compare their funds' performance against various benchmarks. Another example: a manufacturer uses the I-Bridge to analyze statistical quality data.

I-Bridge is designed to simplify development tasks by offering the developer an object model of the data

sources and a set of high level interface components. The components fetch data from UNIX and prepare it for Windows application. By communicating directly with the components instead of the data sources, the developer has a single access method for all types of data sources, including databases, UNIX utilities, and applications that control, analyze or generate data.

Below is the Graphical User Interface. **At right** is a modular representation of a way in which a corporation might use I-Bridge; it can be used to connect different operating systems in an office and to allow workers in one facility to access information from computers in different buildings, cities or different countries. **At lower right,** a group of I-Kinetics computer engineers is working on advancements to the

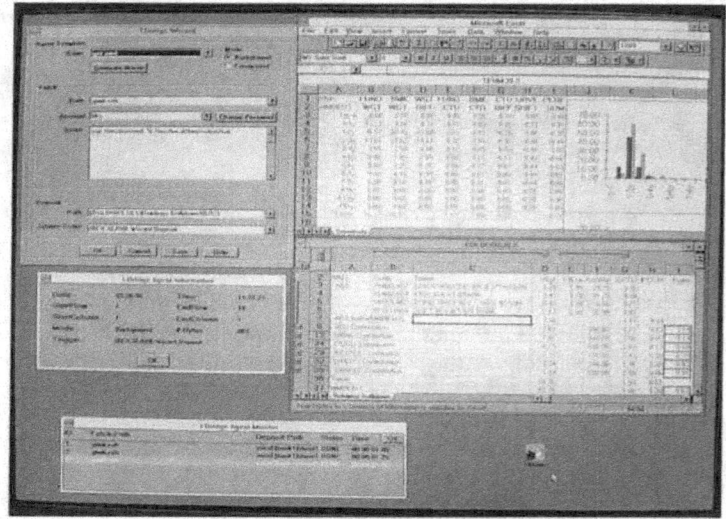

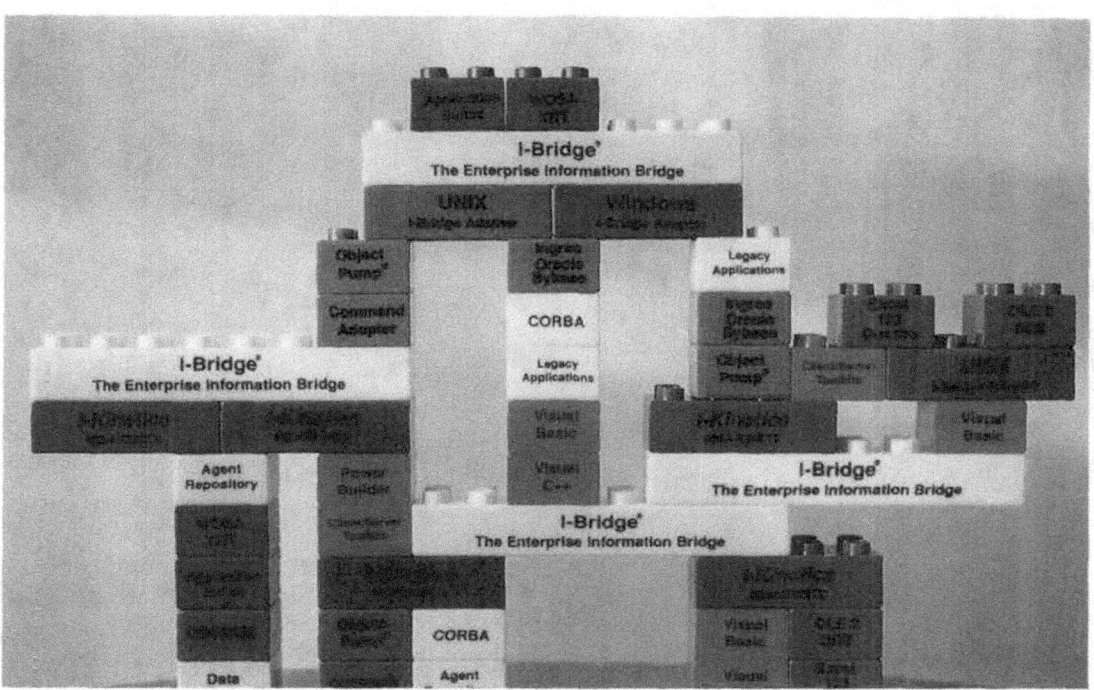

system; at left is company president Dr. Bruce H. Cottman. The I-Bridge family includes object adapters for popular relational databases. Since a large percentage of enterprise data is not in relational databases, I-Bridge also includes an adapter that provides access to these programs and allows Windows applications to invoke UNIX applications and high level language scripts.

I-Kinetics, Inc. was formed by Dr. Cottman, who had earlier co-founded another software development company, Symbiotics, Inc. In 1991, Dr. Cottman led Symbiotic's spinoff of its R&D Division to become I-Kinetics.

™I-Kinetics is a trademark of I-Kinetics, Inc.

®I-Bridge is a registered trademark of I-Kinetics, Inc.

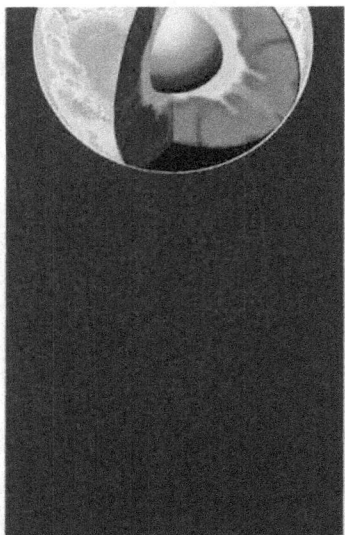

ASTRONOMY SOFTWARE

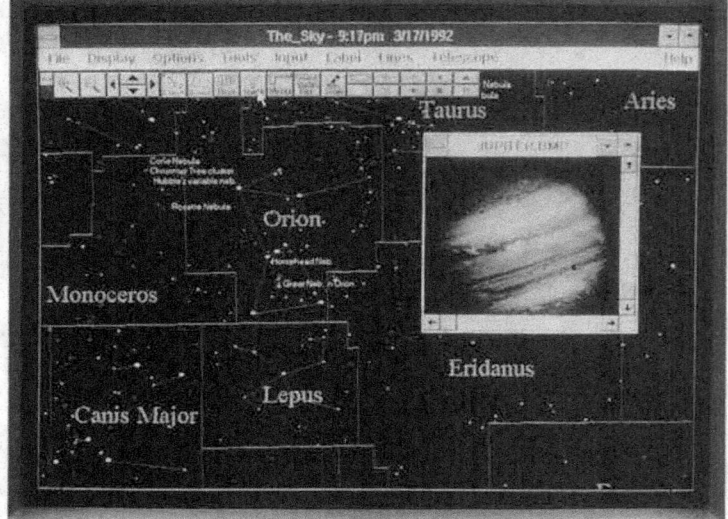

In the accompanying photos are a new astronomy software package called TheSky and a sample chart view generated by the program. TheSky is part of a software family that incorporates NASA astronomy technology developed for the Hubble Space Telescope. The family includes, in addition to TheSky, the SkyPro image processing program and the Remote Astronomy Software for running a telescope and imaging system.

The programs were developed by Software Bisque, Golden, Colorado with NASA input, in particular the 19-

million-object database compiled for the Hubble Space Telescope. With TheSky, amateur and professional astronomers can quickly plot any of the Hubble data onto the computer screen. Dozens

of commands make it easy to locate any area of the sky; the software presents a wide angle view that identifies each constellation and bright object. As shown **above,** the astronomer can use a mouse to zoom in on any zone of the view presented.

The companion SkyPro CCD Astronomy Software is a Windows-based image processing program that also controls CCD (charge coupled device) cameras, transforming raw CCD data into imagery and extracting hidden information. TheSky and SkyPro work in tandem to orchestrate locating, identifying, and acquiring images of deep sky objects.

With TheSky, SkyPro and the Remote Astronomy Software, a user can directly control computer-driven telescopes and CCD cameras through computer serial ports. Telescope support is provided for a number of different types of instruments.

Through a cooperative effort involving Software

Bisque, NASA, The Mount Wilson Observatory Institute, Pasadena, California and the Observatory's Telescopes in Education program, individual astronomers or astronomy classroom groups can remotely operate Mount Wilson's 24-inch telescope and CCD camera, which are fully dedicated to remote astronomy. A Local Serial Port Controller enables TheSky and SkyPro to communicate via modem and phone lines with Mount Wilson; a simple click of the mouse instructs the telescope to point at a selected object contained in TheSky database and SkyPro captures an image of the object by controlling the Observatory's CCD camera. Astronomers can access the telescope by paying a fee based on duration of usage; through a NASA grant, K through 12 school groups can use the telescope free of charge.

ANALYSIS/DESIGN TOOL

Developed by INTERSOLV, Inc., a computer software developer located in Rockville, Maryland, Excelerator II is a new software system described by the company as "smarter and more flexible, the next generation of planning, analysis and design tools." Excelerator II incorporates technology developed by NASA to provide a complete environment for rules-based expert systems.

In addition to built-in support for many popular methodologies, Excelerator II takes advantage of the INTERSOLV LAN Repository's Customizer, which allows a user to construct or tailor any methodology to the user's particular organizational requirements.

In previous versions of the product, INTERSOLV used a procedural language to check the consistency and completeness of the LAN Repository. However, since the main task being executed is comparing patterns, the company decided that a rules-based language was needed. The rules-based language selected was NASA's CLIPS (C Language Integrated Production System), a shell for constructing expert systems. CLIPS was developed by Johnson Space Center to provide high portability, low cost and easy integration with external systems.

INTERSOLV chose CLIPS because the company's technical staff could easily implement the communication to the relational database. The staff was able to expand CLIPS with low level relational database access routines. With this rules-based environment, which the company calls the "INTERSOLV Rules Engine (IRE)," design problems and errors are identified early in development, when they may be corrected quickly and cost effectively.

"The IRE is a key position point in our analysis and design marketing approach," says Shep Bostin, director of Excelerator II product management. "Using CLIPS has enabled us to provide complex verification and transformation routines based on pattern matching that would be an order of magnitude more complex and costly if we had to resort to procedural language."

CLIPS was supplied to INTERSOLV by the Computer Software Management and Information Center (COSMIC)®, NASA's mechanism for making available to industry, academic and government clients computer programs originally developed by government agencies that have secondary applicability (see page 000).

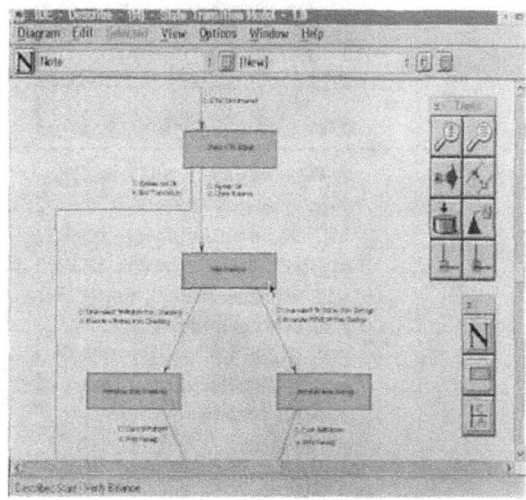

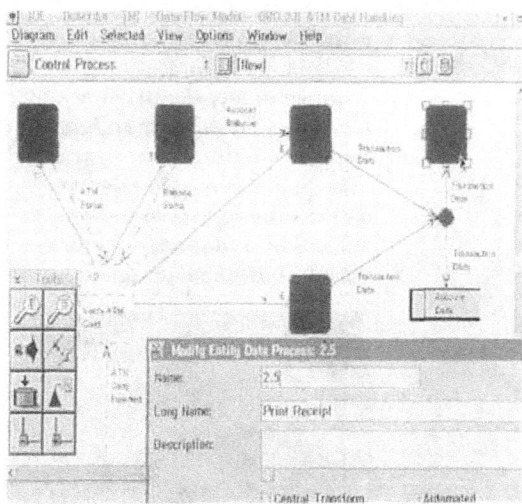

®COSMIC is a registered trademark of the National Aeronautics and Space Administration.

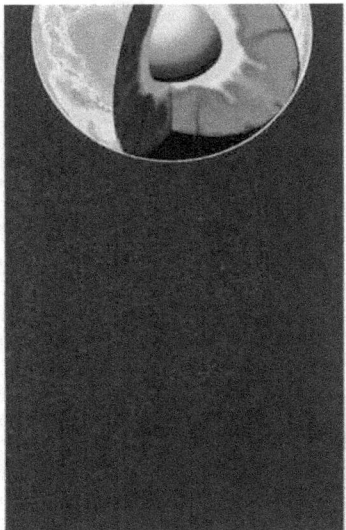

SCIENTIFIC SOFTWARE

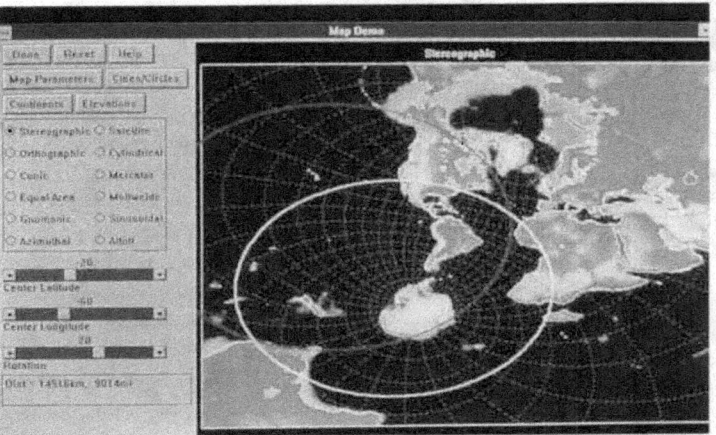

Research Systems, Inc. (RSI), Boulder, Colorado develops and markets software for analysis and visualization of scientific and engineering data. The company's flagship product is IDL® (Interactive Data Language), developed for use with a broad spectrum of computing hardware. More than 20,000 scientists and engineers in laboratories, universities and commercial organizations employ IDL's comprehensive mathematical analysis and graphical display capabilities for research in physics, remote sensing, astronomy, test and measurement, financial visualization and medical imaging.

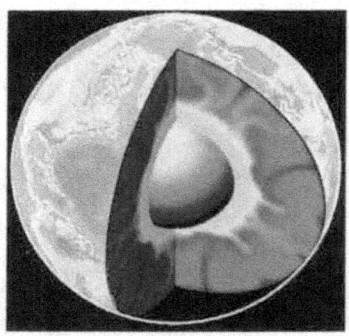

IDL traces its lineage to a predecessor software package known as the Mariner Mars spectral Editor (MMED), developed for NASA's series of Mars flyby spacecraft flown in the late 1960s and early 1970s. The concept embedded in MMED — and the essential element of IDL — was that scientists could test their hypotheses without having to write a custom program every time new data were collected. It provided scientists a general tool that allowed them to investigate their data directly, without the need for programmers or for extensive programming knowledge.

Founded in 1977, RSI began developing IDL that year, using the MMED technology as a departure point. IDL was introduced commercially — for use on a VAX system — in 1982. It has since progressed through four technology generations and the current version runs on all popular workstations and personal computers. NASA, one of RSI's first customers, sponsored development of a Convex version of IDL designed specifically for Goddard Space Flight Center.

The accompanying photos illustrate some IDL applications. **At left** is an IDL 3D rendering showing the magma flow that drives continental plate movements. IDL's use as a mapping tool is shown **above;** this is a stereographic projection in which great circles pass through two user-defined points of interest. **At top right** is an image of the Sun in the x-ray wavelength acquired by a CCD camera/ soft x-ray telescope combination built by Lockheed Solar & Astrophysics Laboratory and flown aboard the Japanese Yohkoh spacecraft. **At center right** is a Magnetic Resonance Imaging slice of a human head displayed by IDL's slicer tool, which allows users to investigate and display 3D volumes of data. **At right** is a screen shot of the surface of Venus displaying two sets of radar

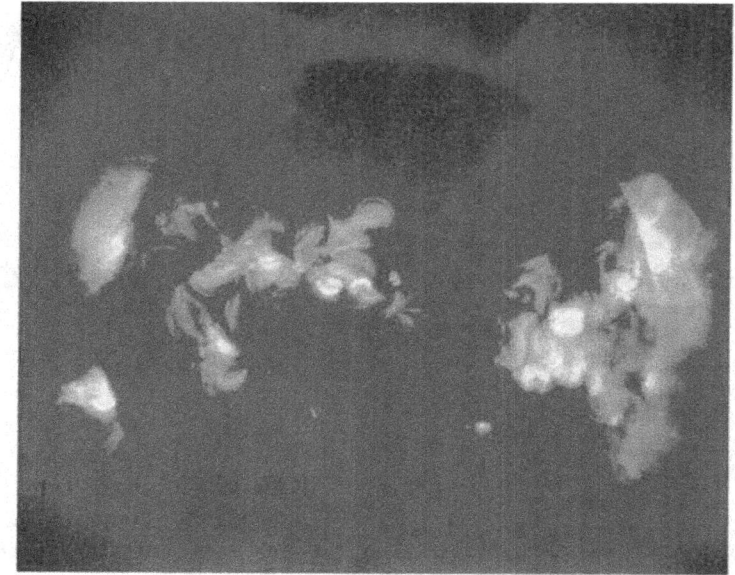

mapping data collected by NASA's Magellan spacecraft; a special IDL capability — known as Show3 — allows researchers to view a single dataset as an image, a contour and a surface.

A recent addition to the RSI software family is ENVI™ (Environment for Visualizing Images), an advanced image processing system for Earth scientists who analyze remotely sensed data. Writ-

ten in IDL, ENVI allows a user to easily read, display, analyze and output any space-acquired or aircraft digital imagery, and to import, manipulate and analyze ground-based and laboratory spectral measurements.

The newest RSI product, released in 1995, is the Visible Human CD, described by the company as "the world's first complete digital reference of photographic images for exploring human anatomy." The body reference product draws on RSI's extensive experience in medical imaging, which dates back to the company's early work in reconstructing multi-dimensional images of the human heart, part of a project that helped scientists build the first practical tomography imaging system.

Developed in cooperation with the National Library of Medicine, the Visible Human CD offers ready access to more than 7,800 physical slices of an anatomical male human, enabling medical professionals, researchers and educators to "navigate" an entire human body on a desktop computer. For efficient storage and retrieval, the image slices are compressed on a single CD ROM.

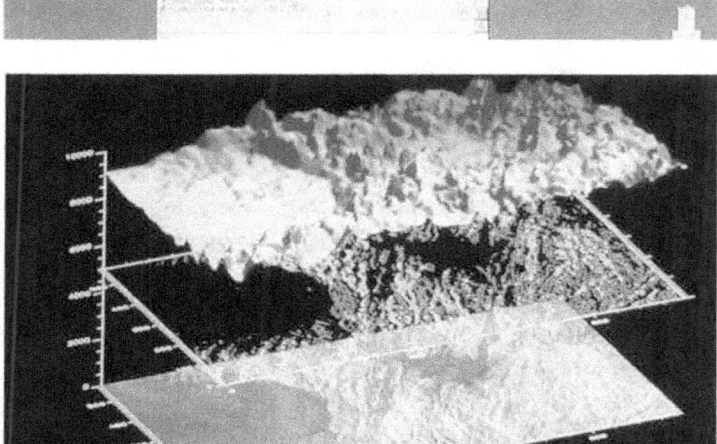

®IDL is a registered trademark of Research Systems, Inc.

™ENVI is a trademark of Research Systems, Inc.

An advanced parts identification system derived from Space Shuttle technology exemplifies spinoff aids to industrial productivity and manufacturing technology

American industry — and government agencies like NASA — are making increasingly greater use of automated parts identification (PID) systems to track the millions of individual parts involved in their operations. Such systems can significantly enhance industrial efficiency when each part is uniquely marked with a machine-readable symbol, or "fingerprint," that provides information about the part and its pedigree.

And therein lies a problem. The growth of automated parts control has been restricted by the fact that a great many tiny or unusually-shaped parts cannot be marked with conventional bar codes such as those used in supermarkets. Some parts — a screw or an electronic diode, for example — are too small to accommodate a printed code label; others must operate in harsh, high temperature environments in which a bar code could not survive.

That was a particular problem for NASA's Marshall Space Flight Center, which was looking for an improved method of identifying and tracking Space Shuttle parts. There was need for a technique of marking parts so that NASA could readily check the specifications, manufacturing history and operational experience data of any single part among the hundreds of thousands in the Shuttle system. And the PID symbol had to be able to withstand operating temperature conditions ranging from minus 250 degrees Fahrenheit to more than 2,000 degrees.

NASA found a possible solution in a technology then under early development by a small business firm, Veritec, Inc., Calabasas, California. Veritec was experimenting with a space efficient binary marking known as the Vericode Symbol™ that could contain a great deal more information than a bar code yet occupy substantially less space; and could easily be marked on a part's surface for permanent identification.

In May 1991, NASA organized a team to further develop the technology and create an advanced PID system that would be targeted specifically at Space Shuttle parts identification but would also be capable of spinoff applications in pharmaceutical, medical, automotive and electronics manufacturing, product distribution and consumer product use. The team included Marshall Space Flight Center, Veritec and Rockwell International, NASA's prime contractor for the Space Shuttle.

The Shuttle-oriented work is still in progress, but the technology has advanced significantly. In cooperation with major firms in several industries, the team has developed marking techniques for many types of parts and a wide range of materials — steel, paper, glass, fabric, ceramics, plastic, metal, etc.

Meanwhile, Veritec introduced the Vericode Symbol to the commercial marketplace in 1994. The company is marketing its VeriSystem™, a complete identification and tracking system for component traceability, improved manufacturing and processing capabilities, and a variety of automated shop floor applications.

The Vericode symbol is a small, checkerboard-like matrix, ranging in size from fractions of an inch to several feet. Each symbol can contain 5 to 100 or more coded characters, which can provide up to 100 times as much information as a bar code in the same or less space. The smallest Vericodes, read with a microscope, are measured in microns (millionths of a meter). Vericodes are scanned by CCD (charge coupled device) electronic cameras that can capture images from a broad range of surfaces. NASA tests have shown that the Vericode Symbol can be read with a high degree of accuracy. The Vericode's direct PID capability also enables bypassing processing steps that introduce human error.

The "compressed symbology" technology clearly has enormous potential for efficiency and cost savings. The applications and methodology vary widely from one possible use to another, but one major example serves to underline the system's utility: labeling automotive parts.

Today, when a manufacturer discovers a defect in a particular batch of auto parts, the company has little information on which autos are specifically affected. Therefore, it becomes necessary to recall thousands of vehicles for inspections, when only a few might really be affected. A Vericode Symbol on a defective part might include such details as the manufacturer, serial number, the lot number of the parent material, design changes, special processing to which the part was subjected — everything needed to determine, accurately and automatically, the extent of the recall needed, which might be a couple of hundred cars instead of tens of thousands. The savings in time, convenience and money are obvious. There is similar savings potential in an ever-broadening spectrum of applications ranging from marking pharmaceutical vials to tracing instances of food poisoning by marking live hogs.

TMVericode Symbol and VeriSystem are trademarks of Ventec, Inc.

SUPERCONDUCTING MATERIALS

Superconductivity is the phenomenon wherein the electrical resistance of a metal disappears when the metal is cooled. Superconductivity occurs in a variety of metals, but only when they are cooled to extremely low temperatures, near absolute zero.

Ever since the discovery of superconductivity in 1911, researchers have sought to raise the temperature at which superconductivity occurs, because that would make possible exciting applications of resistance-free electricity long thought impractical in light of the cooling requirement — for example, supercomputers with speed and capacity several orders of magnitude beyond today's capabilities; frictionless superspeed trains levitated by magnetic force; transmission lines that suffer no power loss through resistance; and supermagnets that could significantly enhance the medical diagnostic technology of magnetic resonance imaging (MRI).

In 1986, researchers achieved a breakthrough by identifying a new class of high temperature superconducting (HTS) materials. The "high temperature" is high only in the relative sense; these oxide materials remain superconductive above 30 degrees Kelvin, which corresponds to 243 degrees below zero on the more familiar Centigrade scale. However, some HTS materials superconduct at temperatures more than 50 degrees higher than the initial applications of superconductivity. That makes them easier to cool and easier to use, opening up a wide range of applications to take advantage of the superconductor's special properties: zero resistance, magnetic field exclusion, low noise and extremely low power loss for high frequency electronics.

With the advent of HTS materials, superconductors have begun to emerge from the laboratory and appear in practical applications.

A pioneer in this explosively advancing technology is Superconducting Technologies, Inc. (STI), Santa Barbara, California. STI's basic products are high quality thin films, sold to manufacturers of communications systems and military systems, who use them for in-house fabrication and component development. STI also produces custom circuits and components — resonators, filters, oscillators, microwave mixers, etc. — for manufacturers who prefer to have the superconducting fabrication and assembly done by specialists. STI uses thallium, the highest temperature material for making HTS. Thallium remains superconductive at temperatures above 77 degrees Kelvin and can be cooled to working temperature by a liquid nitrogen system instead of the

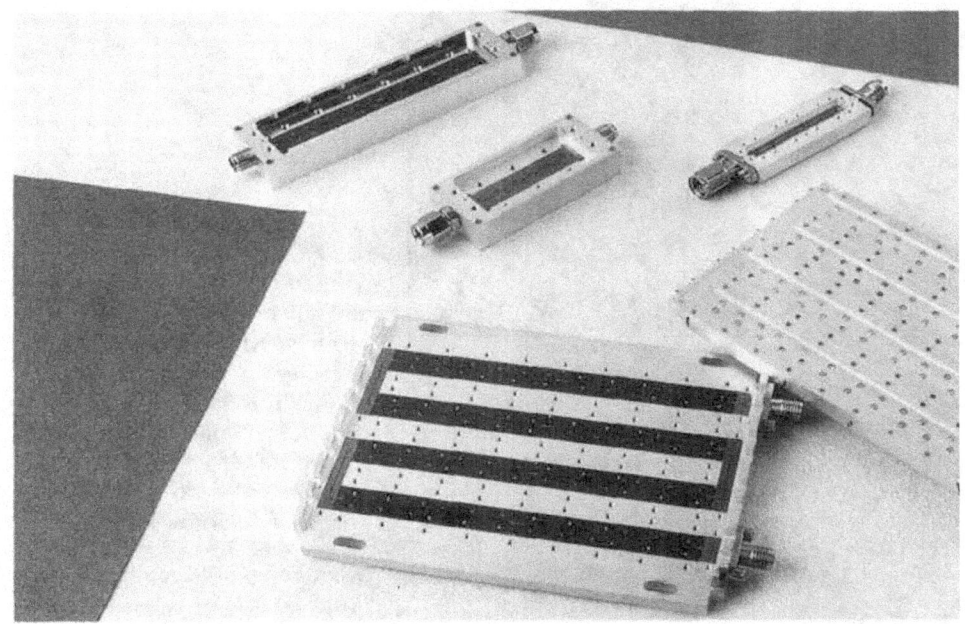

more difficult and more expensive helium method.

At left, HTS is being produced by a laser ablation system. **Below,** an STI scientist displays the result, a three-inch wafer of HTS material. **At right** is a representative application: microwave circuits used in radars to reduce interference; the dark brown strips are made of HTS material. **At lower right** is a closed cycle stirling cooler used to cool HTS subsystems to 77 degrees Kelvin; it has a one-quart capacity and the same electricity as an ordinary light bulb.

In addition to producing films, circuits and components, STI provides a wide range of foundry services, allowing access to HTS technology by company engineers who have not had previous experience in HTS. A number of government agencies and several large

companies are using STI's foundry facility and personnel expertise to design and develop HTS circuits and components for specific applications.

STI credits NASA with a technological assist in the company's climb to a leadership position in the development and commercialization of superconducting products. Founded in 1986, STI has worked with two NASA centers — Lewis Research Center and Jet Propulsion Laboratory (JPL) — and advanced its own technology by adapting to NASA requirements. Says Joseph M. Madden, manager of sales and applications, "In fabricating antenna circuits for Lewis and other devices for JPL, we refined our standard production recipe for circuits. This now enables us to use the same standard production recipe for our production work."

SMALL BUSINESS INNOVATIONS

A steadily growing source of aerospace spinoff applications is the Small Business Innovation Research (SBIR) program, which was established by Congress in 1982 with dual objectives: to increase participation by small businesses in federal R&D activities, and to stimulate conversion of government funded research into commercial applications.

NASA is one of 11 technology-generating agencies of the federal government participating, each administering its own programs independently under policy guidelines set by the Small Business Administration.

Among recent examples of successful SBIR commercialization is the PER-Force hand-controller shown **at right,** a force-reflecting system that manipulates robots or objects by "feel." Originally developed for the International Space Station under a Johnson Space Center SBIR contract, the handcontroller is produced by Cybernet Systems Corporation, Ann Arbor, Michigan. Headed by president Heidi Jacobus and her husband, Dr. Charles Jacobus, Cybernet began life in 1988 as a cottage industry and has grown to a work force of 25; the company has worked on 16 SBIR projects and still does most of its business with government agencies, but is making gradual inroads in the commercial market.

Cybernet's original NASA contract called for development of a teleoperation device with force feedback that would allow a space station astronaut to position an external robot arm. When the arm encountered an outside force, the robot control stick would represent that force by making motors built into the joystick push or pull or twist to represent the reality of the outside world. The intent was to improve teleoperator performance by providing the sense of feel.

In the course of developing the NASA system, Cybernet found it necessary to go beyond representing only real forces, and to include virtual reality representations. "Since then," says Dr. Charles Jacobus, who is vice president and technical director, "we've discovered that the flexibility incorporated into the control system for building virtual force fields has opened up a whole new dimension in man-machine interface for complex visualization applications." The device is particularly well-suited for teleoperator environments where direct viewing is limited, for example, in murky underwater environments or reduced light areas such as underground excavations.

The commercially available PER-Force Handcontroller is a small, back-driveable robot handle that moves in six degrees of freedom, including 3-D positions (x, y, z) and three attitudes (roll, pitch and yaw). The operator uses the motorized handle to precisely position

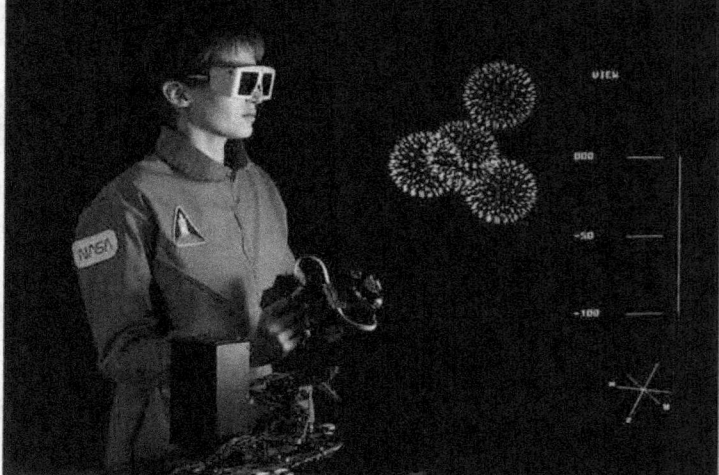

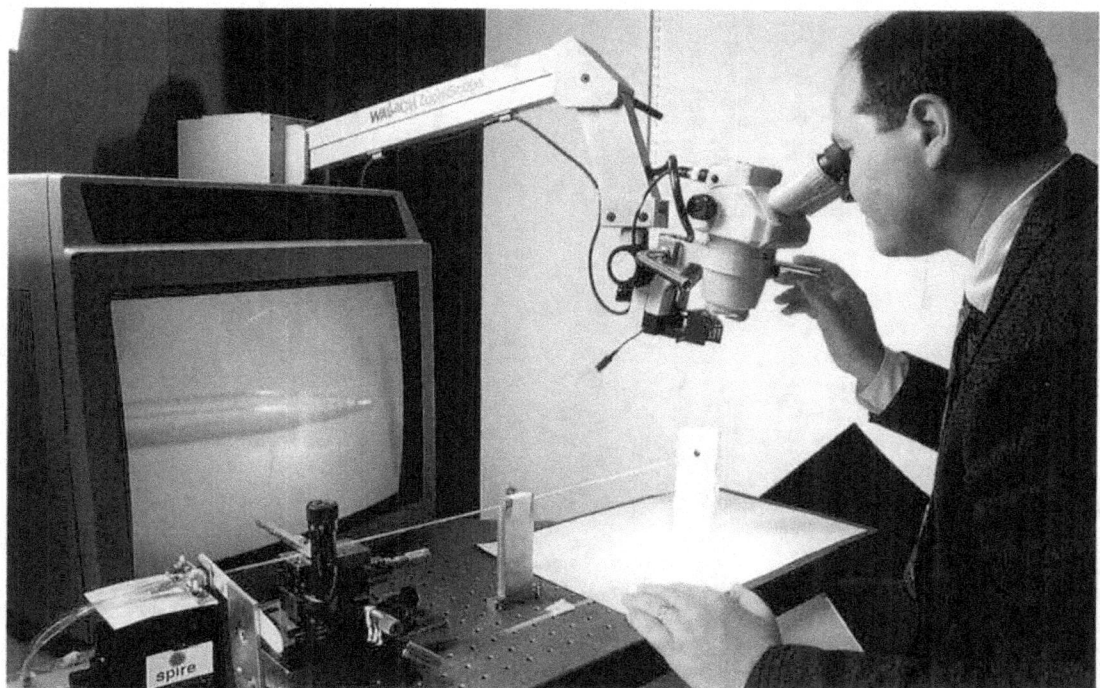

robots and graphically displayed objects at a given location and tool angle. Among current applications, it is being used for molecular modeling in metallurgy applications, in satellite docking research, and in research on military unmanned ground vehicles. Real and virtual reality forces are simulated by a 3-D molecular modeling software package that calculates the interacting forces among attracting and repelling molecules; those forces are represented through the robot stick.

Cybernet has used the original SBIR work as a departure point for a whole family of force reflective devices in the telepresence and virtual reality fields, and has developed and patented a general interface architecture for integrating visual and tactile displays.

Another example of SBIR success is the commercialization of high power diode laser arrays, developed under Langley Research Center SBIR funding, by Spire Corporation, Bedford, Massachusetts. Langley's principal interest in diode laser arrays was their potential for use in LIDAR (light detection and ranging) devices in windshear detection and warning systems. A Langley developed windshear system employs an optical laser telescope to send light beams ahead of an aircraft into a storm; measurement of the energy reflected back to the telescope from particles in the storm provides an indication of windshear.

The NASA contract with Spire called for development of a high power semiconductor diode laser module to pump solid state laser rods or slabs. This type of module, as a replacement for the normally used flashlamp, offers a number of advantages: highly efficient optical pumping, very small size, low drive voltage requirements, and extended operational lifetimes.

Spire Corporation successfully developed the laser array for NASA and used the NASA work as a development base for a commercial line of diode laser arrays for industrial and other uses. **Above,** Dr. Joel S. Schuman investigates a high power pulsed diode laser at Tufts University School of Medicine, Medford, Massachusetts. The laser delivers its light through a thin fiber optic cable. Among applications are laser sclerectomy, an operation to reduce glaucoma-caused eye pressure. The products can also be used in industrial cutting and drilling and in military applications. Spire Corporation's Dr. Kurt J. Linden, manager of laser product development, reports that the high efficiency of the products has brought rapidly increasing demand for diode laser pumps.

(Continued)

113

INDUSTRIAL
PRODUCTIVITY

Langley Research Center has been engaged for more than a decade in development of advanced polymers, used in foams, fibers, adhesives, composites and coatings for a variety of space applications. The Center is among the world leaders in developing these chemical compounds, and Langley sponsored advances have contributed significantly to improving the U.S. competitive posture in polymer production and application.

Langley, however, is an R&D organization, so it looks to industry to commercialize the Center-invented polymers, identify commercial applications in and beyond the aerospace field, and develop cost effective ways of producing them. This is accomplished by partnerships with industry firms under Small Business Innovation Research (SBIR) contracts.

An example is the collaboration of Langley with Triton Systems, Inc. (TSI), Chelmsford, Massachusetts. TSI is an innovative high technology company specializing in R&D and technology transfer leading to product development. The company develops niche products in high performance polymers, advanced ceramics, metal matrix composites and flexible manufacturing processes.

A major need for the International Space Station and future long-duration space structures is a polymer resistant to atomic oxygen (A/O), which exists in low Earth orbit and causes corrosion of spacecraft surfaces. Langley researchers invented a promising high performance polymer known as PAEBI to meet that need. Under an SBIR contract, TSI took the PAEBI polymer from laboratory scale to pilot scale, identified several niche market opportunities, and formed a partnership with a leading polymer film processing firm. TSI was granted a license for commercial development of the PAEBI-based polymer, now known as AORIMIDE®

(Atomic Oxygen Resistant Imide), and the company also developed the processing techniques needed to coat long runs of high quality continuous films of AORIMIDE.

TSI subsequently received three other SBIR contracts, each geared toward a different AORIMIDE-based product. TSI now offers the product in four forms: AORIMIDE polymer in pound quantities for commercial uses; AORIMIDE free standing films for space systems and for Earth use in electronics and electrical insulation applications; ultrahigh performance polymer threads based on AORIMIDE chemistry for space applications; and AORIMIDE co-polymer material for use as membranes in elevated temperature separation applications (it is in use by a major membrane manufacturer as a liquid separation medium). **Above,** Triton president Ross Haghighat displays one of his

products, sheet polymer rolled into a core; other products including powder, liquid, and polymer sheet, are shown **at left below.**

Another small business that has worked effectively as a Langley partner is Imitec, Inc., Schenectady, New York. As the company name suggests, Imitec focuses its efforts on polyimides, polymers that can tolerate high temperatures without deforming. Imitec has received several SBIR contracts to modify, characterize and commercialize Langley-invented polyimides.

Imitec developed its own processes to produce the Langley materials, processes that make possible fiber quality material, film quality material and moldable material. Under a Langley SBIR contract, Imitec teamed with Barcel Wire and Cable, Irving, California in a successful project involving development of polymer pelletizing and extrusion technology and its application to manufacture of insulated aircraft cable. Imitec is also collaborating with AlliedSignal, Hoosick Falls, New York on use of NASA polyimides in flexible circuitry.

Above, Imitec chemist Betty Chung is making the material in a reactor; **at right** is a sampling of products made from the material, including high temperature gears, polymer in sheets,

polymer in bead form, and temperature resistant light-weight brick or tile forms.

Imitec has installed a reactor, centrifuge, film casting equipment, drying ovens and additional electric power to continue commercialization of the NASA polymers. The company has also developed processes to manufacture monomers and installed a pilot reactor to continue development. Imitec is supplying developmental quantities of powder or liquid poly (amic acid) to such customers as Lockheed Martin, Northrop Grumman and ICI-Fiberite for aerospace composites and optics; Cytec Industries, The Boeing Company and 3M Corporation for use in adhesives; IBM and Motorola for electronic systems; Ford Motor Company

and Delco-Remy GM for automobile applications. Imitec is also working with Lockheed Martin and Rockwell International on using Langley invented polymers in the NASA X-33 launch vehicle development program.

(Continued)

*AORIMIDE is a registered trademark of Triton Systems, Inc.

NASA's Langley Research Center was looking for a way to improve the process of inspecting aging aircraft without taking the planes apart. They found it in a novel concept — originated by Richard Albert **(right)** and developed by his company, Digiray® Corporation, San Ramon, California — called reverse geometry x-radiography (RGX)®. RGX employs a combination of television-like scanning and digital data acquisition to produce real-time x-rays with film quality.

Since the discovery of x-rays in 1895, radiographers have employed a small point source and a large detection area. This conventional configuration allows some degradation of image quality as the detection medium registers the secondary radiation (x-ray scatter) produced in the object being bombarded by radiation.

RGX reverses the conventional configuration. The object to be x-rayed is placed adjacent to a large source whose raster scan is picked up by a distant detector. In this approach, much of the unwanted scatter generated in the object is absorbed by the intermediate air before it can reach the remote detector. The result is improved image clarity and better inspection throughput, with

four times the contrast sensitivity of conventional systems.

Under two Small Business Innovation Research (SBIR) contracts, Digiray teamed with Langley researchers of the Center's Nondestructive Evaluation Sciences Branch (NESB) to develop a portable RGX system (portability is a necessity for field inspection of aircraft), and to upgrade the x-ray energy of the system so that thicker parts could be x-rayed. The SBIR work involved miniaturization of the system's sensors so they could be inserted into internal aircraft structures, such as hard-to-get-at corners and crevices. The project gave Langley a device that can be used not only for aircraft inspections, but for assessing damage growth in materials, for supporting tests to show how structures behave under stress, and for monitoring changes in solid

rocket fuel over time.

The Digiray project was part of a still-ongoing seven year Langley program that is exploring ways to conduct thorough examinations of aircraft without disassembling them by means of advanced inspection techniques, including ultrasound and thermography in addition to x-ray.

Digiray further developed the technology under separate arrangements with the U.S. Air Force, the Department of Energy's CEBAF (Continuous Beam Accelerator Facility) and Lawrence Livermore National Laboratory.

The RGX produces x-ray images by means of a scintillating crystal detector in which a number of tiny crystals light up when excited by x-ray beams, creating an image that is computer processed into a digital x-ray. The image can then be enhanced by standard image processing tools, such as

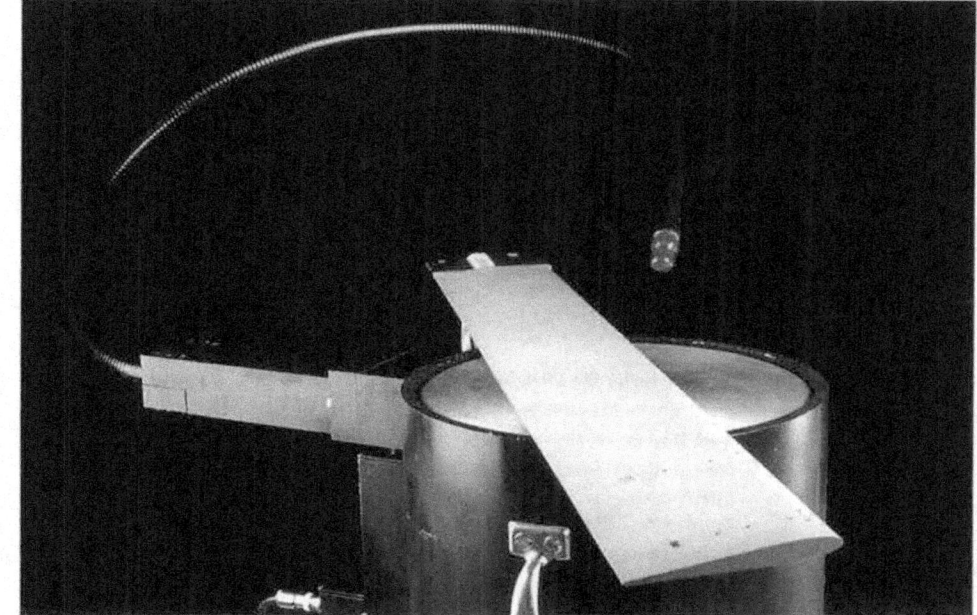

averaging, filtering, image subtraction and edge enhancement. **At right,** a helicopter tail rotor is undergoing inspection by Digiray to find delamination and stress cracks; **below,** a Digiray image shows corrosion in an airplane wing assembly.

The feasibility of using the RGX system for computed tomography, which provides cross-sectional x-ray images, has been demonstrated. With the RGX, an entire family of cross-sectional images can be obtained during a single revolution of an object.

The RGX is being used by the Air Force to x-ray parts of fighter aircraft, and by McDonnell Douglas Corporation in a comparison study of RGX versus ultrasound inspection of composite aircraft structures. The portability of the system has led to applications in areas other than aerospace, for example, inspection of pipes corroding under insulation at oil refineries, which has been

identified as a cause of serious refinery fires; **at right** is an image showing thinning in the walls of a refinery pipe section. Mobil Oil Corporation, Exxon and Shell Oil are conducting tests of the Digiray system.

A related product that emerged from the Langley work is the company's patented RGL (reverse geometry laminography) system that provides layer-by-layer x-ray viewing with only one exposure. A demonstration of the RGL showed the system's ability to image both sides of a quarter — the Washington head and the eagle tail — with a single exposure. The clarity of the

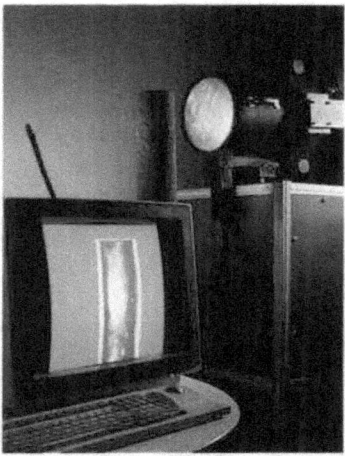

imagery suggests that the system can enhance current imaging techniques in such areas as mammography, cardiac imaging, brain surgery and orthopedics.

(Continued)

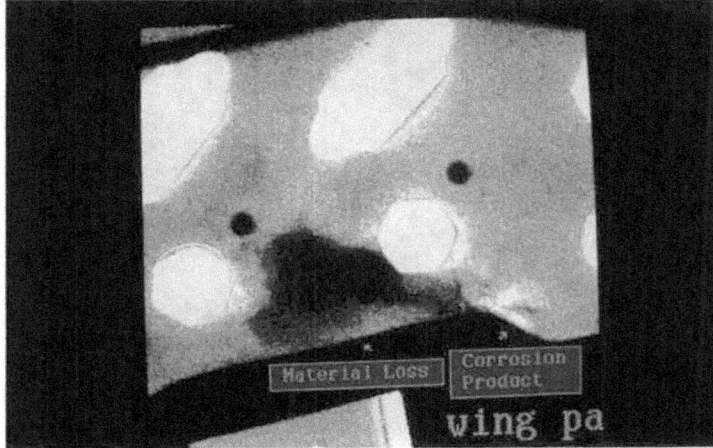

Continually looking for new ways to shave weight and improve the structural integrity of Space Shuttle components, NASA awarded several Small Business Innovation Research (SBIR) contracts to Nichols Research Corporation (NRC), Huntsville, Alabama for development of an advanced system to weld segments of the 154-foot-long Space Shuttle External Tank. NRC successfully developed such a system and commercialized the technology with the marketing of a device known as Wire Pilot™.

Essentially a small robot, the Wire Pilot **(below)** is a programmable, motorized three-axis manipulator for precise positioning of filler wire relative to the welding torch and workpiece. Designed to automate the placement of filler wire during precision automatic and robotic welding, the wire manipulator is housed in a compact, rugged, nine-inch package that accepts standard wire feeder guide tips. A hand-held unit **(right)** enables control of the filler wire entry angle, the force against the workpiece, and

the offset from the weld centerline. The controller electronics are shown **at lower right.**

A miniature load cell built into the wire guide tip assembly senses the force between the workpiece and the filler wire, and closed loop control drives the motors to keep this force constant, thus maintaining a constant force between the filler wire and the workpiece regardless of automatic voltage control (AVC) adjustments or reactions to condition changes. The load cell output can also be used as an AVC feedback

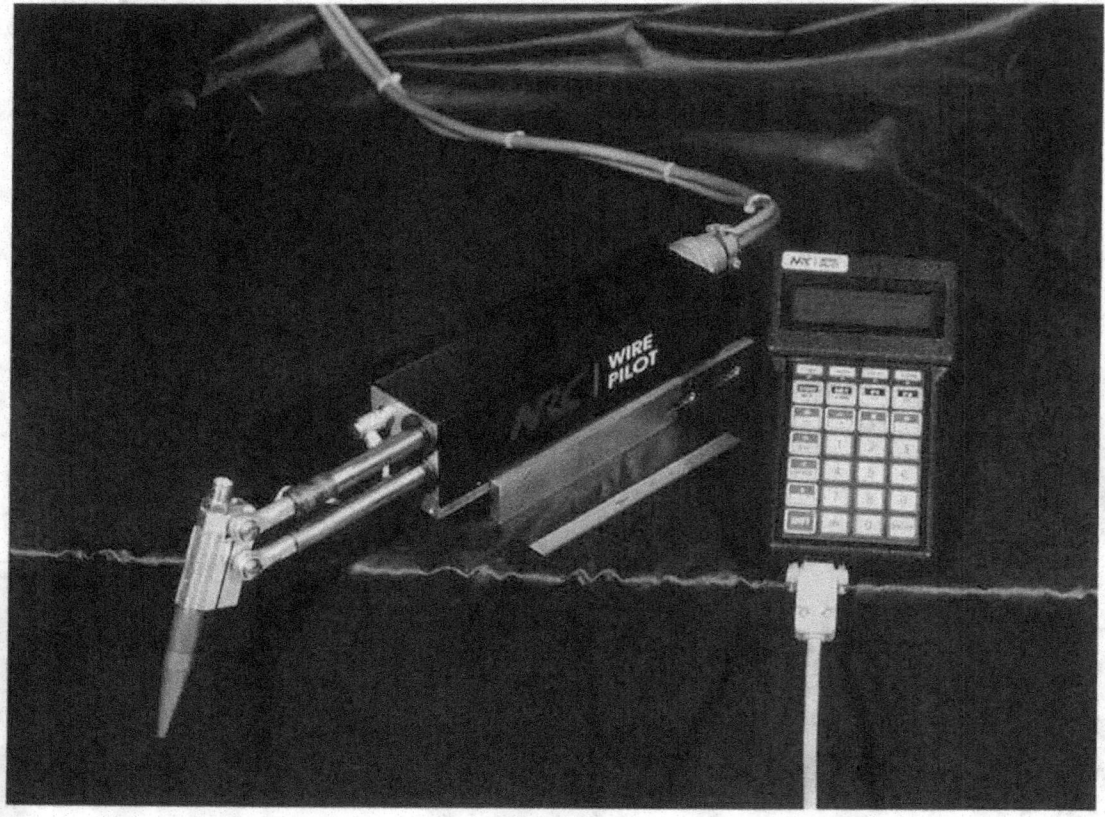

signal to provide constant torch-to-wire distance.

The Wire Pilot employs a unique configuration of three linear drives to achieve high precision three-axis coordinated motion. The three axes of motion are the X axis (side-to-side), the Y axis (in/out), and the Z axis (angle change). The Z axis rotates about an operator-defined toolpoint that is programmed to the location where the wire enters the weld pool. This prevents angle changes from affecting the force on the wire pressure load cell.

Because operator or equipment malfunctions can cause the unit to be crashed into the workpiece, a two-stage crash protection system is built into the device. For slight crashes, a spring on the upper drive rod flexes, allowing the tip to rotate without sustaining damage. In more severe crashes, the entire unit slides backward up to one inch on its base plate without damage. It is

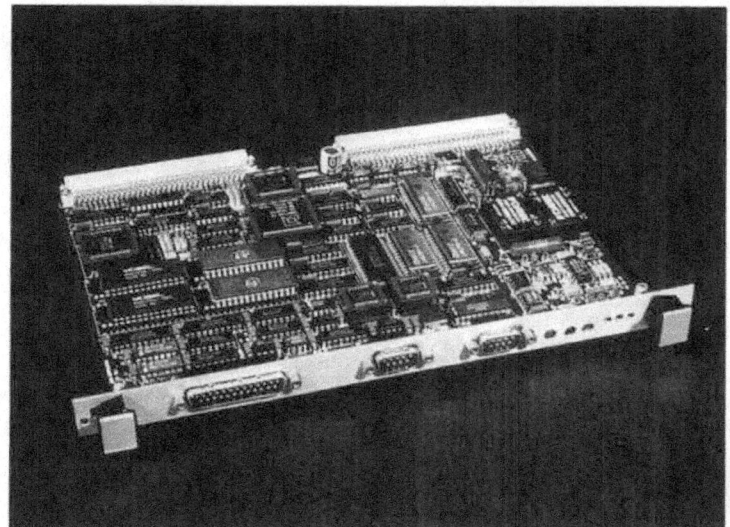

spring loaded to return to its normal position.

The Wire Pilot was developed as a stand-alone system or as part of a VME-based control system. The stand-alone system consists of the wire manipulator, the controller and an operator's pendant. The controller board can be integrated directly into VME-based systems.

Marshall Space Flight Center managed the SBIR contracts and engineers/technicians of Marshall's Materials Processor Branch assisted NRC throughout the development of the system.

™Wire Pilot is a trademark of Nichols Research Corporation.

119

INDUSTRIAL PRODUCTIVITY

STRAIN GAGE

In August 1994, an eight-legged mobile robot named Dante II, developed for NASA by Carnegie Mellon Robotic Institute, descended into the depths of Alaska's Mount Spurr. An active volcano that had erupted three times in 1992, Mount Spurr was considered too dangerous for human exploration, so Dante II, equipped with a multitude of sensors and imaging devices, was sent to rappel down the inside wall of the crater and communicate its findings. The volcano expedition was an important scientific investigation and, at the same

time, a demonstration of technology applicable to future planetary rovers.

Among a number of innovations incorporated in Dante II was a unique strain

gage application developed especially for the project by HITEC Corporation, Westford, Massachusetts. Carnegie Mellon selected HITEC for the job because the company specializes in experimental or otherwise unusual installations of strain gages and thermocouples. The demanding job assigned HITEC was to develop a strain gage-based solution for measuring bending forces on the legs of Dante II; the sensor was intended to provide a warning to the robot's human controllers, who could shut Dante down if, for example, one of the robot's legs got caught up in the rocky terrain and the kickout load on the leg exceeded design limitations.

The addition of the sensor was a late decision and HITEC had only a month to conduct the extensive research required and produce a solution. HITEC engineers managed it. They came up with a sensor system consisting of four strain gages, two of them measuring bending in tension at the surface strain on one side, while two pages on the opposite side measured compression bending.

The work on Dante benefitted HITEC because it expanded the company's technological capability. HITEC senior applications engineer Robert Magee reports that "it allowed HITEC to further develop its

technology for taking existing customer-supplied components and turning them into accurate and reliable transducers. The work on the robot explorer has furthered our experience in strain gage-based sensor applications." HITEC uses similar techniques providing strain gage services in creating transducers out of such components as "Indy" racing car suspension pushrods **(left)** NASCAR suspension components and components used in motion control. Generally, HITEC applies sensors to measure strain, stress and loads on automotive, gas turbine and structural components. **Above** is the first U.S.-manufactured bobsled for the U.S. Bobsled Team; HITEC installed sensors on the four skate runner shoes to measure vibration forces.

FIBER OPTIC ATTENUATORS

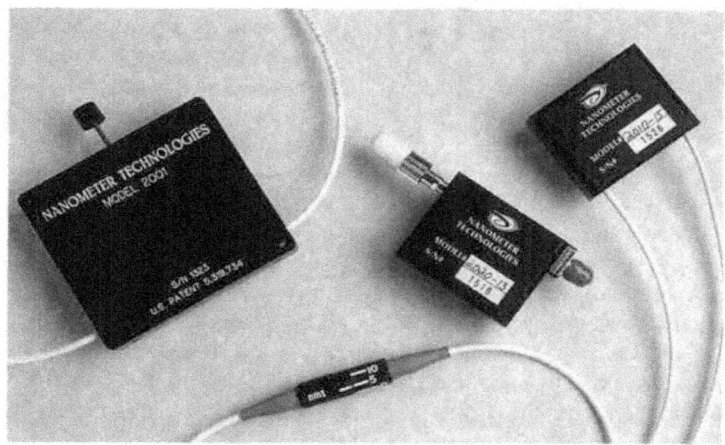

One of the routes by which many spinoff products are brought to the commercial marketplace is personnel technology transfer, wherein an individual acquires a particular skill or technological know-how in the course of his work as an employee of NASA or a NASA contractor, then moves to a new job or a new industry and applies that skill or knowledge to a useful commercial product. An example is the experience of Mike Buzzetti **(below),** now president and owner of Nanometer Technologies, Santa Clarita, California, who designed a fiber optic attenuator while working at NASA's Jet Propulsion Laboratory (JPL), subsequently commercialized the device and founded a company to produce it.

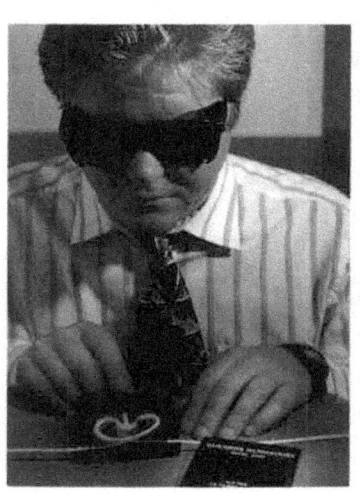

In 1990, Buzzetti was a contract technician at JPL engaged in research on a very low noise Fiber Optic Distribution Assembly for use in NASA's Deep Space Network. In the course of this research, the JPL team discovered that all of the fiber optic attenuators tested showed instability due to variations in temperature and shock, and had too much back-reflection (reflected light in a fiber system). Test results were unacceptable.

Buzzetti then designed a new, adjustable fiber optic attenuator that utilizes bending loss to reduce the strength of the light transmitted along it. The attenuator functions without introducing measurable back reflection or insertion loss, and it is relatively insensitive to vibration and changes in temperature.

Buzzetti applied for and received a patent for the device, negotiated an exclusive licensing agreement with California Institute of Technology, JPL's parent organization, and started Nanometer Technologies to manufacture and market the product, which has applications in cable television, telephone networks, other signal distribution networks, and laboratory instrumentation.

Above are three units of Nanometer's Model 2000 Series single mode fiber optic attenuators. By midyear 1995, the company had sold more than 1,000 units.

VIBRATION-RESISTANT FASTENER

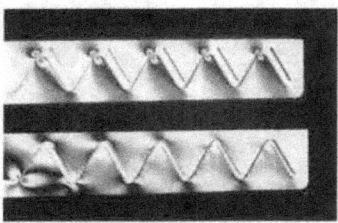

Each of the three main engines of the Space Shuttle generates 400,000 pounds of thrust and, of course, terrific vibration. In developing the engines, NASA looked for a special type of fastener, one that could withstand the great shock and vibration of a Shuttle launch without loosening. That was a very big order, but it got bigger: NASA also wanted a 15-cycle reuse capability for the fastener.

Those demanding requirements are met — and exceeded — by the Spiralock® internal thread form type of fastener, manufactured by Detroit Tool Industries, Madison Heights, Michigan. NASA

tests of the superfastener showed that the Spiralock did not back off or loosen even when subjected to vibrations 10 times greater than the specified requirement. And the fastener easily topped the 15-cycle requirement; it was found to have no loss of clamping power after 50 cycles. Every Shuttle engine built incorporates more than 750 Spiralock fasteners.

The key Spiralock feature is a unique 30 degree wedge ramp, outlined in blue **at left.** When clamp load is applied during assembly, the Spiralock thread form locks the standard male fastener in place by drawing the crests of the male thread tightly against the wedge ramp. That "wedge-locking" makes Spiralock exceptionally resistant to transverse vibration, the major cause of thread loosening, and substantially reduces the potential of fatigue failure.

To better visualize the unique load distribution properties of Spiralock, Detroit Tool conducted a photoelastic study of conventional and Spiralock threads. The photo **above** illustrates the difference between Spiralock (upper strip) and a standard thread form (lower strip); in the latter the load is concentrated on the first thread and in the Spiralock the load is distributed evenly across the length of the fastener.

A user testimonial to Spiralock's effectiveness is supplied by Philips Broadband Networks, Inc. (PBN), Manlius, New York. A manufacturer of broadband radio frequency and fiber optic networks, Philips had long used a special, high quality fastener of another make due to the large number of insertions and withdrawals necessary in certain of the company's products. The fastener provided a high level of joint integrity but it was costly and difficult to assemble.

Pete Strong of the PBN engineering department read in *NASA Tech Briefs* (see page 137) about the use of Spiralock thread form in space applications. He brought the article to the attention of the Mechanical Engineering Group, which decided to perform comparative strength and insertion/withdrawal testing of their regularly used fastener and Spiralock over a wide temperature range. The performance results for both fasteners were similar, and both far exceeded PBN's customers' requirements. PBN was able to switch to Spiralock in most of its products, eliminating a troublesome installation problem and gaining a bonus in cost reduction on certain product lines where Spiralock fasteners contributed to easier assembly.

*Spiralock is a registered trademark of Detroit Tool Industries.

VOLTAGE-CONTROLLED OSCILLATOR

Integrated Component Systems, Inc. (ICS), Coconut Creek, Florida designs and manufacturers synthesizers and oscillators used in wireless systems, modems, test equipment and related products. ICS products range in size from inch-square surface mount units to full-size boards for frequency-hopping communications systems.

ICS president Eliot Fenton describes his firm as "an innovative high technology company with very limited financial resources, always looking for outside information to assist in the development process." He found a valuable source in *NASA Tech Briefs,* a monthly publication

used as a problem solving tool by more than 200,000 government and industry subscribers (see page 137). A *Tech Briefs* article provided information incorporated in an ICS custom design.

One of the company's customers wanted a synthesizer with wide phase modulation characteristics, yet low noise. The problem was that the two requirements are inherently incompatible in a single loop design. An ICS engineer suggested an approach, but Fenton wanted confirmation of the validity of the method. The *Tech Briefs* article described research at Jet Propulsion Laboratory (JPL) resulting in a modified configuration for a phase-locked angle modulator that makes it possible to design the filters in the modulating portion of the circuit independently of the filter in the phase-locked

loop portion; applied to a phase-locked oscillator, it offers superior phase noise performance.

"The article gave my engineer valuable insight into how the process works," says Fenton, "and substantiated our method as viable for wideband phase modulation." The technology is incorporated in the ICS series of phase-locked loop synthesizers. Fenton (hand on console) talks with a company engineer **above; at left** are a pair of representative ICS oscillators.

TECHNOLOGY TRANSFER

A description of the mechanisms employed to encourage and facilitate practical application of new technologies developed in the course of NASA activities

A nationwide technology transfer network seeks to broaden and accelerate secondary use of NASA-developed technology

Because they are challenging and technologically demanding, NASA programs generate a great wealth of advanced technology. This bank of technology is a national asset that can be reused to develop new products and processes, to the benefit of the U.S. economy in new companies, new jobs, and the resulting contribution to the Gross National Product.

Such "spinoff" applications do not happen automatically. It takes a well-organized effort to put the technology to work in new ways and to reap thereby a dividend on the national investment in aerospace research.

NASA accomplishes that end by means of its Technology Transfer Program, which employs a variety of mechanisms to stimulate the transfer of aerospace technology to other sectors of the economy. The program is managed by the Commercial Development and Technology Transfer Division of NASA's Office of Space Access and Technology. Headquartered in Washington, D.C., the division coordinates the activities of technology transfer offices located throughout the United States.

Among the most important mechanisms are the technology transfer offices at NASA's 10 field centers. These offices differ somewhat from center to center, but generally their jobs involve promoting transfer and commercialization of technology that has significant potential for secondary use. Representative of this type of activity is the Technology Programs and Commercialization Office (DE-TPO) at Kennedy Space Center (KSC).

For more efficient utilization of resources, KSC has combined a number of technology-related functions in this single office, some of them not directly part of the transfer process. DE-TPO, therefore, is not exclusively a technology transfer mechanism; it is an umbrella group that embraces the work of the traditional NASA center Technology Transfer Office (see page 128), some new transfer-related mechanisms, together with management of technology development projects in support of the KSC mission, projects that may or may not have commercialization potential.

An example of a KSC mission-related requirement that resulted in a commercial product is the Universal Signal Conditioning Amplifier (USCA), developed under KSC's Dual Use Program. For its own operations, KSC had need for a system that could universally connect all types of sensors and transducers to current and proposed data acquisition systems. Working with I-NET, an engineering support contractor, KSC researchers developed the USCA. Since the amplifier clearly has broad applicability in industrial operations, the system was further developed under a dual use agreement by NASA, the Florida Technological Research and Development Authority (TRDA) in cooperation with a private sector partner, Loral Test and Information systems, and several Florida colleges and universities. The USCA was commercialized and the first units manufactured in 1995. Another technology transfer activity at DE-TPO is KSC's participation in the NASA Commercial Technology Mission Electronic Network, which provides U.S. industry electronic access to information on available NASA technologies, facilities and expertise and serves as an electronic marketplace for facilitating communications/

transactions between NASA and U.S. public/private sector organizations. KSC's Commercial Technology Home Page offers continually-updated information on KSC commercialization opportunities, patent and licensing information, alerts on "hot" technologies available for transfer, and details of KSC facilities available for commercial use.

One other technology transfer activity at KSC is the Outreach program, which seeks to transfer the benefits of NASA's unique technical expertise to industrial firms in the Southeast United States (KSC has joined with Marshall Space Flight Center and Stennis Space Center in a Southeast Regional Outreach Alliance; KSC supports the Alliance through the Florida TRDA and County Economic Development Councils throughout the state).

These Councils periodically visit industrial facilities in Florida to make known the NASA technology transfer process. Council members invite Florida businesses to submit their technical problems to the Alliance in the form of a Technology Transfer Agreement, which states the company's problem and requests NASA assistance in solving it.

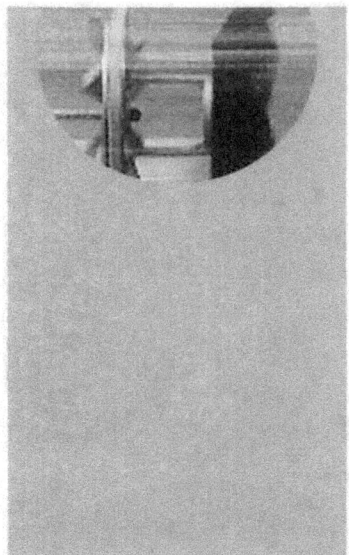

TECHNOLOGY TRANSFER OFFICERS

An important element among the NASA mechanisms for promoting technology transfer is the Technology Transfer Officer, or TTO. TTOs are technology transfer experts based at each of NASA's 10 field centers and one specialized facility who serve as regional managers for the Technology Transfer Program.

Representative of the group is Jim Aliberti, TTO at Kennedy Space Center (KSC), shown **below** with KSC associate Dan Culbertson. At KSC,

the TTO is part of a larger office known as the Technology Programs and Commercialization Office, or DE-TPO (see page 126). The TTO in this instance is known officially as Head of Technology Transfer and Commercialization.

The TTO's basic responsibility is to stay abreast of research and development activities at his center that have significant potential for generating transferrable knowledge. He assures that the center's professional people identify, document and report new technology developed in the center's laboratories and, together with other center personnel, he monitors the center's contracts to see that NASA contractors similarly document and report new technology, as required by law. This tech-

nology, whether developed in-house or by contractors, becomes part of the NASA bank of technology available for transfer.

To advise potential users of the technology's availability, the TTO evaluates and processes selected new technology reports for announcement in NASA publications and other dissemination media.

A KSC mechanism for announcing developments with commercial potential is the "Technology Opportunity," a single page flyer containing a description and photo of a technology developed initially for KSC's use, and an invitation to private sector firms to transfer the technology to the commercial marketplace by licensing, technical consultation, cooperative development with

NASA, or development, manufacture and marketing by the company.

An example is the Surface Defect Analyzer shown **at left.** It was developed by KSC as an alternative method for in-field evaluation of surface flaws, defects and damage on critical surfaces of the Space Shuttle and its ground support equipment. The Technology Opportunity sought a contractor to refine the technology and broaden the system's capabilities for possible applications in medicine and dentistry, laboratories, and the aerospace, automotive, precision tooling and appliance manufacturing industries.

Other examples include the Hydrogen Fire Detector Calibration Unit **(below),** commercially applicable as a device for ensuring confidence in ultraviolet flame detectors used on oil rigs, in refineries, aircraft hangars, etc. to detect hydrogen

flames; and the Turbine Brush Pipe Cleaner **(top),** which offers enhanced cleaning action and possible elimination of chlorofluorocarbons in such applications as maintenance of water supply lines, cleaning dairy lines, controlling corrosion in steam lines, and in-service cleaning of food processing and

pharmaceutical equipment.

The TTO helps promote technology transfer internally at his center and to the public in general, particularly to businesses that might be able to use NASA technology productively. At KSC and other centers, this is accomplished by notices through print and electronic media, and by showcasing selected technologies at seminars, conferences and symposia. **Above,** KSC's Dave Makufka mans a technology transfer booth at the 1995 Space Congress in Cocoa Beach, Florida.

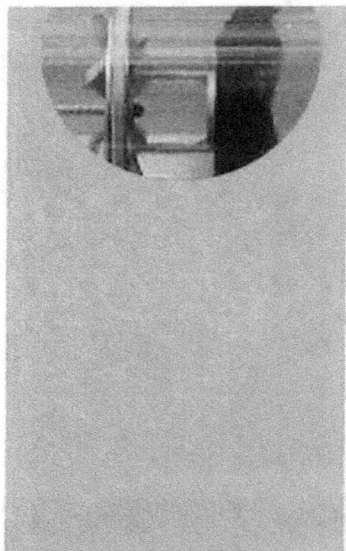

TECHNOLOGY TRANSFER NETWORK

To promote technology transfer, NASA operates a number of user assistance centers whose job is to provide information retrieval services and technical help to industrial and government users.

Intended to meet the technological needs of American industry and boost U.S. international competitiveness, the National Technology Transfer Network is composed of a National Technology Transfer Center (see page 132) and six Regional Technology Transfer Centers (RTTCs). The RTTCs are geographically located to provide an equal distribution of services throughout the country. The

regional deployment of the centers and their alignment with the Federal Laboratory Consortium allows the RTTCs to work closely with federal, state and local programs in serving the technol-

ogy related needs of business and industry.

The RTTCs provide value-added services to meet the needs of clients, including

- Information services: computerized searches of federal technology databases and other technology sources.

- Technical services: assessment of technology requirements, analysis of technology applications, and engineering reports.

- Commercialization services: technology brokering, business analyses and venture capital sourcing.

Other elements of the National Technology Transfer Network include:

- Federal agency technology transfer programs and activities.

- State and local agencies and their programs, such as technology centers and business/technical assistance centers.

- Business and industry consortia, associations and communities. **The photo at left** illustrates an effective way in which the RTTCs seek out contacts in the business community, through booth contacts at trade shows.

A typical RTTC provides a wide range of technology management services, including information retrieval **(above),** technical analyses and assessments, market intelligence, product enhancement and applications development. The center conducts searches in a wide variety of commercially available government and proprietary databases to provide information on

state-of-the-art developments in a particular field, individuals and organizations conducting relevant R&D, patents and licenses, and technologies that can be redeployed or reengineered to meet a client's needs.

After searching the databases, the RTTC investigates noncomputerized sources of materials, such as traditional print information. Experts from NASA field centers, federal laboratories, universities and industrial organizations are contacted to locate additional information.

The RTTCs support NASA's educational objectives by providing teachers access to information generated by NASA programs. The center makes available scientific and technological materials related to research and technological developments in the areas of life sciences, physical science, astronomy, energy, Earth resources, geology, mathematics and astronautics. **Above,** an RTTC staffer at a Teachers Resource Center helps a teacher search for materials to enliven lesson plans. **At left** is a sampling of materials available to teachers and other clients of the RTTCs.

(Continued)

The hub of the National Technology Transfer Network is the National Technology Transfer Center (NTTC), located at Wheeling Jesuit College, Wheeling, West Virginia. The NTTC serves as a clearinghouse for federal technology transfer, linking U.S. firms with federal agencies and laboratories, the RTTCs, and state/local agencies.

The center operates, in the interest of enhanced U.S. competitiveness, a "gateway service," toll-free telephone access to a full federal technology database and indexing system. By calling a 1-800 number, U.S. companies can access the federal laboratory system in search of technologies and research data that can assist them in developing their businesses.

The NTTC also provides training and educational services to government and industry to develop the skills essential to effective technology transfer. Additionally, the center conducts outreach and promotional activities to improve private sector awareness of technology transfer opportunities.

One such activity is NTTC's national electronic bulletin service for both the public and the private sectors of the U.S. technology transfer community. The free service includes notices of upcoming technology transfer related meetings, announcements of new technologies, problem query/answer communications, success stories and posting of opportunities.

Support of all the elements of the National Technology Transfer Network is provided by the Technology Transfer Office at the Center for Aero-Space Information (CASI). This office executes a wide variety of tasks, among them maintenance of the subscription list for and mailout of Technical Support Packages (TSPs), which provide details of new technologies available for more than 70 percent of the listings published in *NASA Tech Briefs* (see page 137). The mailout of TSPs involves a reproduction effort of more than 1.8 million pages annually. CASI is also responsible for responding to requests for information, an activity that entails processing of some 40,000 letters and other inquiries and mailout of some 300,000 documents a year. The office additionally serves as a "central call-in" facility, channeling information and technical assistance seekers to the proper NASA technology transfer and commercialization organization or other appropriate agencies.

Pictured **at left** is Walter M. Heiland, manager of the CASI Technology Transfer Office. **At left above,** technology associate Jennifer Munro is reviewing photographs for the *Spinoff* publication. **At lower left,** Mindy Murdza (standing) and Maria Zimmerman are preparing correspondence and **above** Lenora Parris (standing) and Diane Odachowski are processing TSP packages requested by clients, as are Sharon Jiles and Audrey Skinner **below. At right,** Kelen White (foreground) and Mary Crum are handling data entries.

The CASI Technology Transfer Office is also responsible for research, analysis and other work associated with this annual *Spinoff* volume, for distribution of technology transfer publications, for retrieval of technical information and referral of highly detailed technical requests to appropriate offices; for developing reference and biographical data; and for public relations activities connected with media, industry and trade show interest in technology transfer matters and commercialization.

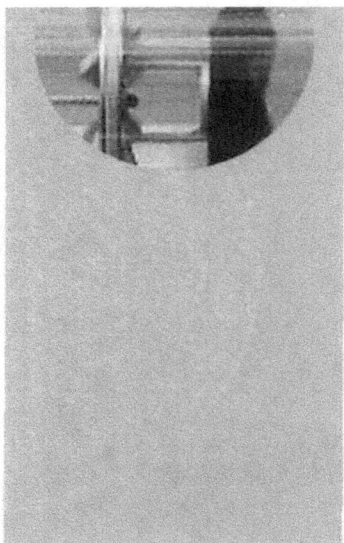

TECHNOLOGY APPLICATIONS

Applications engineering projects are efforts to solve significant public sector or industrial problems through redesign or reengineering of existing technology. They originate in various ways. Some stem from requests for assistance from other government agencies; others are generated by NASA technologists who perceive possible solutions to problems by adapting NASA technology to the need. NASA also employs an applications team composed of scientists and engineers representing different areas of expertise, who identify problems, submit them to NASA centers for review, then assist the centers in adapting solutions.

An example of an ongoing applications engineering project is an effort at Langley Research Center to develop a Fetal Heart Rate Monitor for monitoring problem pregnancies.

Some 20 percent of the three million annual pregnancies in the U.S. require periodic non-stress evaluations that rely on fetal heart rate changes as a barometer of the baby's health. The heart beat changes in response to influences such as drug and alcohol abuse, smoking, ane-mia, high blood pressure, diabetes or infection.

Conceived by Donald A. Baker, M.D., of Spokane, Washington and developed by Dr. Allan J. Zuckerwar, a research scientist at Langley, the portable Fetal Heart Rate Monitor is an acoustically-based passive instrument that permits a high-risk patient to perform the non-stress test at home on a daily basis (the acoustic approach, in contrast to the usual ultrasound technique, does not involve creation and injection of energy into the patient and her fetus).

The monitor **(below)** is a cummerbund-type sash into which is embedded seven piezoelectric polymer film pressure sensors that, together with signal processing, enable separation of fetal heart sounds from the mother's. The patient dons the sash, activates the associated desktop computer and, upon completion of the 20-minute test, awaits instructions displayed on the screen. The system can be available on a rental basis and the cost of each test is expected to be substantially less than the current in-hospital or in-clinic test.

Clinical testing is in progress at Eastern Virginia Medical School, Norfolk, Virginia and Tarzana (California) Regional Medical Center. An exclusive license has been offered to Veatronics, Charlotte, North Carolina to manufacture and market the system. Commercialization forecasts project a sales potential of $100 million a year in the U.S. alone.

Another example is Langley's development of an Oxidation Catalyst that can convert toxic carbon

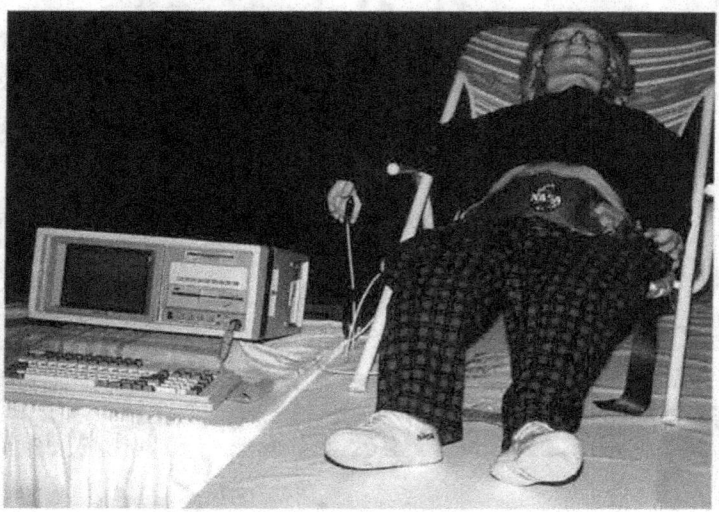

monoxide and formaldehyde fumes into harmless carbon dioxide and oxygen. The catalyst can be used in a broad range of applications, notably in air purification systems for homes and office buildings.

A team led by Dr. Billy Upchurch and David R. Schryer, both Langley chemists, developed the catalyst for use in a space-based laser that uses carbon dioxide to help generate its beam. In the process, some carbon dioxide molecules are decomposed into carbon monoxide and oxygen. Without replenishment, the loss of carbon dioxide would eventually cause failure of the laser. Therefore, NASA sought a way to regenerate the carbon dioxide by catalytically recombining the carbon monoxide and oxygen, a way of providing a long term supply of carbon dioxide for a laser aboard an environmental satellite that would have to operate for at least five years.

Upchurch and Schryer spent parts of 10 years look-

ing for a way to create carbon dioxide by oxidizing another substance, such as carbon monoxide. In 1992 they found the proper catalyst, a balanced combination of tin alloy and platinum plus a "promoter" to enhance the catalytic activity, and successfully demonstrated the catalyst in a laser system. They continued development, aiming at practical Earth applications, and in 1995 demonstrated a home-use catalyst shown **above** (the white disc is an uncoated, commercially available ceramic catalyst support, the black disc is the Oxidation Catalyst with the tin/platinum material added).

The catalyst looks like a filter but it is not; it does not simply absorb harmful gases, it oxidizes them, and it permits breathable gases to pass through unchanged. Rochester (New York) Gas and Electric is developing the catalyst in a home air circulation system for removing carbon monoxide and formaldehyde (used in furniture and carpet manufacture)

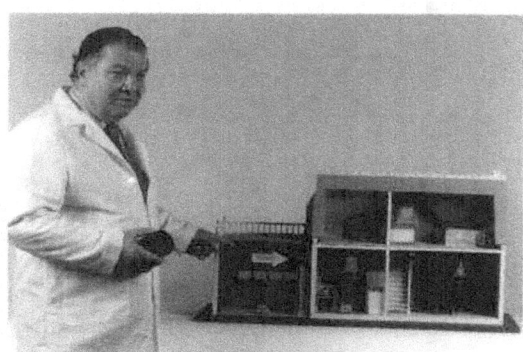

from indoor air; **at right above,** Upchurch is explaining that application. Other applications include gas masks, submarine ventilation systems, and catalytic converters for cold-start emission control of automotive exhausts. The basic technique can be modified to use other catalytic materials and oxidize other gases for a wider range of applications.

In the course of its varied activities, NASA makes extensive use of computer programs, as do other technology generating agencies of the federal government. To meet their software requirements, these agencies have of necessity developed many new types of programs.

These programs constitute a valuable resource that is available for reuse. Much of the software is directly applicable to secondary use with little or no modification; most of it can be adapted to special purposes at a cost less than that of developing a new program.

Therefore, American businesses can save time and money by taking advantage

of a special NASA service that offers programs capable of being adapted to new ones. NASA's mechanism for making the software available to businesses and other clients is the Computer Software Management and Infor-

mation Center (COSMIC)®.

Located at the University of Georgia, COSMIC gets a continuous flow of government-developed software and identifies those programs that can be adapted to secondary usage. The center's library contains more than 1,200 programs for such purposes as structural analysis, artificial intelligence, computational fluid dynamics, thermal analysis, image processing, project management and a great variety of other functions. COSMIC customers can purchase a program for a fraction of the original cost and get a return many times the investment, even when the cost of adapting a program to a new use is included.

An example of how this service aids industry is the use of a COSMIC program by American Airlines Knowledge Systems, a group of software engineers who develop systems utilizing artificial intelligence technologies for such applications as aircraft maintenance, aircraft scheduling, fare administration, flight operations and computer operations.

The group employs a variety of knowledge-based and expert systems, among them a NASA-developed, COSMIC-supplied software package known as CLIPS (C Language Integrated Production System). CLIPS is used for three purposes: as a rapid prototyping tool, to develop production prototypes, and to

develop production applications. An example of the latter is CLIPS' use in the airline process known as "Hub SIAAshing," a knowledge-based system that recommends contingency plans for American Airlines System Operations Control (SOC) during inclement weather or other airport/airways disruptions where severe schedule reductions must be made.

Prior to the introduction of Hub SIAAshing, SOC **(left)** used printouts of flight operations transactions to locate contingency plans manually, used colored markers to highlight candidate plans, weeded through the possibilities and marked them manually. This labor intensive process has been automated by Hub SIAAshing, which integrates the CLIPS shell with more traditional technologies.

With Hub SIAAshing, the search for appropriate patterns is accomplished programmatically, saving time and allowing SOC coordinators to handle more difficult situations. Hub SIAAshing locates the candidates for schedule reduction and assimilates much of the information necessary to facilitate educated decision-making, thus minimizing the negative impact in situations where it is impossible to operate all scheduled flights.

*COSMIC is a registered trademark of the National Aeronautics and Space Administration.

An essential measure in promoting greater use of NASA technology is letting potential users know what technologies are available for transfer. This is accomplished primarily by the publication *NASA Tech Briefs.*

The National Aeronautics and Space Act requires that NASA contractors furnish written reports containing technical information about inventions, improvements and innovations developed in the course of work for NASA. These reports provide the input for *NASA Tech Briefs.* Issued monthly, the free publication is a current awareness medium and problem solving tool for more than 200,000 government and industry readers.

Each issue contains information on newly developed products and processes, advances in basic and applied research, improvements in shop and laboratory techniques, new sources of technical data and computer programs, and other innovations originating at NASA field centers or at the facilities of NASA contractors.

An example of how *Tech Briefs* inspires secondary application of NASA technology is the experience of a small firm — Performance Extremes R&D, Norman, Oklahoma — producing motorcycle parts. The company was started on the idea of making parts out of carbon/epoxy to reduce weight, but the first fabrication attempts were "pitiful," according to president Larry Ortega, pictured **below.**

A friend introduced Ortega to *Tech Briefs* and called attention to an article that detailed a vacuum bagging process for forming composite parts, wherein a part is placed in a vacuum bag,

drawn under vacuum with a pump, then heat-cured in an oven. The company bought a vacuum pump and started using the process. It was a partial answer to the company's problem; it produced high quality parts but a new difficulty cropped up: the parts warped after cooling.

Ortega and his partners went to the University of Oklahoma library, searched a number of back issues of *Tech Briefs* and found another answer: a modification of the vacuum bagging process that made for more even heat transfer. On a third occasion, Performance Extremes found yet another useful technique in *Tech Briefs* and adapted it to advantage. Larry Ortega no longer has to go to the library to search *Tech Briefs;* he is a subscriber and an appreciative one. **Above** is an example of a finished Performance Extremes part, a motorcycle muffler.

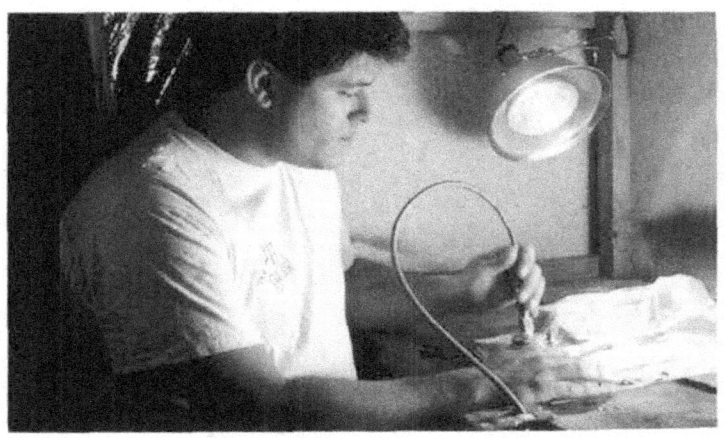

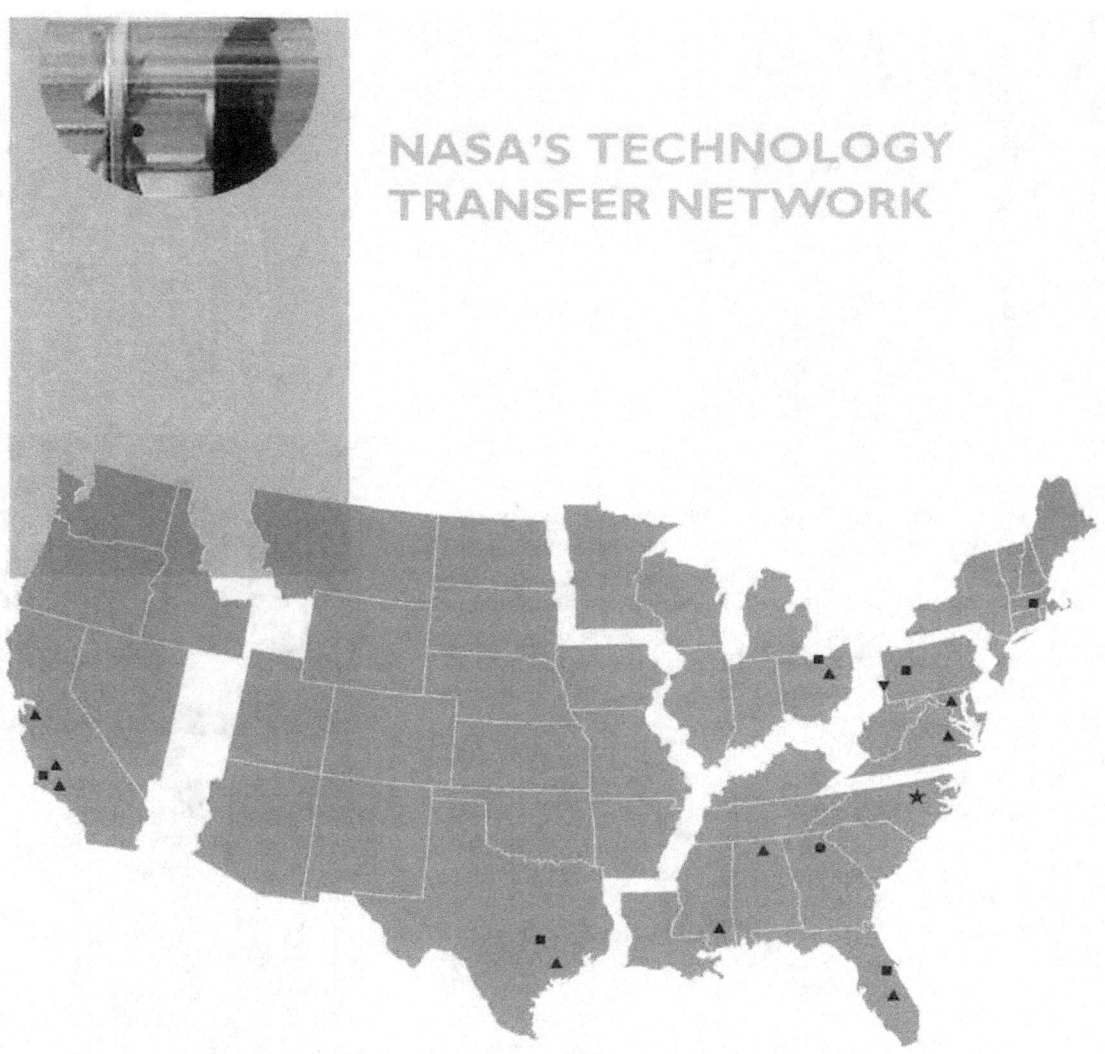

The NASA system of technology transfer personnel and facilities extends from coast to coast. For specific information concerning the activities described below, contact the appropriate technology transfer personnel at the addresses listed, or address inquiries to the Manager, Technology Transfer Office, Center for AeroSpace Information, 800 Elkridge Landing Road, Linthicum Heights, Maryland 21090.

▲ *Field Center Technology Utilization Officers:* manage center participation in regional technology transfer activities.

▼ *National Technology Transfer Center:* national information, referral and commercialization service for NASA and other government laboratories.

■ *Regional Technology Transfer Centers:* information, technical and commercialization services.

● *The Computer Software Management and Information Center (COSMIC):* offers government-developed computer programs adaptable to secondary use.

★ *Application Team:* assists agencies and private institutions in applying aerospace technology to solution of public problems.

Ames Research Center
National Aeronautics and
Space Administration
Moffett Field,
California 94035
Director, Office of
Commercial Technology:
Syed Z. Shariq, Ph.D
Phone: (415) 604-0753

**Goddard Space
Flight Center**
National Aeronautics and
Space Administration
Greenbelt, Maryland 20771
Chief, Office of Commercial
Programs:
George Alcorn, Ph.D.
Phone: (301) 286-5810

**Lyndon B. Johnson
Space Center**
National Aeronautics and
Space Administration
Houston, Texas 77058
Director,
Technology Transfer and
Commercialization Office:
Henry Davis
Phone: (713) 483-0474

**John F. Kennedy
Space Center**
National Aeronautics and
Space Administration
Kennedy Space Center,
Florida 32899
Technology Utilization
Officer: *James A. Aliberti*
Phone: (407) 867-3017

Langley Research Center
National Aeronautics and
Space Administration
Hampton, Virginia 23681-0001
Technology Transfer Team
Acting Branch Chief,
Technology Utilization
Charles P. Blankenship
Phone: (804) 864-6005

Lewis Research Center
National Aeronautics and
Space Administration
21000 Brookpark Road
Cleveland, Ohio 44135
Chief, Office of Commercial
Development:
Ann Heyward
Phone: (216) 433-3484

**George C. Marshall
Space Flight Center**
National Aeronautics and
Space Administration
Marshall Space Flight Center,
Alabama 35812
Manager, Technology
Transfer Office:
Harry G. Craft, Jr.
Phone: (205) 544-5418

Jet Propulsion Laboratory
4800 Oak Grove Drive
Pasadena, California 91109
Technology Transfer Office
Manager: *Merle McKenzie*
Phone: (818) 354-2577

**NASA Management
Office—JPL**
4800 Oak Grove Drive
Pasadena, California 91109
Technology Utilization
Officer: *Arif Husain*
Phone: (818) 354-4862

**John C. Stennis
Space Center**
Mississippi 39529
Technology Transfer Officer:
Anne Johnson
Phone: (601) 688-3757

**Dryden Flight
Research Facility**
National Aeronautics and
Space Administration
Post Office Box 273
Edwards,
California 93523-0273
Chief, Technology and
Commercialization
Office: *Lee Duke*
(805) 258-3802

1-800-472-6785 You will be
connected to the RTTC in
your geographical region.

FAR WEST
Technology Transfer Center
University of Southern
California
3716 South Hope Street,
Suite 200
Los Angeles, California 90007
Robert Stark, director
Phone: (213) 743-2353
(800) 642-2872 (toll-free US)

NORTHEAST
*Center for Technology
Commercialization*
1400 Computer Drive
Westborough,
Massachusetts 01581
*William Gasko, Ph.D.,
director*
Phone: (508) 870-0042

MIDWEST
Great Lakes Industrial Technology Center
25000 Great Northern Corp. Ctr., Suite 260
Cleveland, Ohio 44070-5331
Christopher Coburn, director
Phone: (216) 734-0094

SOUTHEAST
Southern Technology Application Center
University of Florida
College of Engineering
Box 24
One Progress Boulevard
Alachua, Florida 32615-9987
J. Ronald Thornton, director
Phone: (904) 462-3913

MID-CONTINENT
Texas Engineering Extension Service
Texas A&M University System
301 Tarrow Street
College Station, Texas 77843
Gary Sera, director
Phone: (409) 845-8762

MID-ATLANTIC
University of Pittsburgh
823 William Pitt Union
Pittsburgh, Pennsylvania 15260
Lani Hummel, director
Phone: (412) 648-7000
(800) 257-2725 (toll-free US)

● COMPUTER SOFTWARE MANAGEMENT AND INFORMATION CENTER

COSMIC
382 E. Broad Street
University of Georgia
Athens, Georgia 30602
Tim Peacock, director
Phone: (706) 542-3265

★ TECHNOLOGY APPLICATION TEAM

Research Triangle Institute
Post Office Box 12194
Research Triangle Park, North Carolina 27709
Doris Rouse, Ph.D., director
Phone: (919) 541-6980

▼ NATIONAL TECHNOLOGY TRANSFER CENTER

Wheeling Jesuit College
Wheeling, West Virginia 26003
Lee W. Rivers, director
Phone: (304) 243-2455
(800) 678-6882

NASA CENTER FOR AEROSPACE INFORMATION

Technology Transfer Office
800 Elkridge Landing Road
Linthicum Heights, Maryland 21090
Walter Heiland, manager
Phone: (301) 621-0241

Spinoff Team:
Project Manager:
Walter Heiland

Senior Technology Associate:
Jane Lynn-Jones

Technology Associate:
Mindy Murdza

Graphic Services:
Another Color, Inc.

Photography:
Kevin Wilson